FIGHTING THE GOOD FIGHT
MINING'S BATTLE FOR SURVIVAL IN THE AMERICAN WEST

VOLUME 4

PINE NUT PRESS
Minden, Nevada

Copyright © 2018 Susan Lee Parkhurst

All rights reserved.

ISBN: 1978450087
ISBN-13: 978-1978450080

Library of Congress Control Number:
2017960238

The Dave W. Parkhurst Mining Writing Collection
was compiled, edited and designed
by Susan Lee Parkhurst
and produced by Pine Nut Press.

Note to Reader

The content of this book is the work of late mining writer/consultant Dave W. Parkhurst (deceased, 1993). The articles contained herein are reprinted from the *California Mining Journal* (now the *International California Mining Journal*, or *ICMJ*) with permission from the *ICMJ* publisher. No substantive changes have been made to the text in the process of digitizing the material for publication in this book, and no liability is assumed for any inaccuracies that may have occurred in the process of converting the printed magazine articles to the digital format, or from the digital format to the printed book. Further, the content reflects the state of the mining industry and its technologies and processes, as well as the knowledge base that existed, as of the time the articles were published in the *CMJ*. No guarantee is made as to the accuracy or current relevance of any of the content.

To all those who

continue

to fight the good fight

for

individual liberty

Contents

Foreword vii
Editor's Preface ix
Editor's Acknowledgements x
Introduction 3

PART I: THE U.S. GOVERNMENT VS. MINING
1   Economic Policies and the Mining Industry.. 7
2   Great Economic Sacrifice, the 13
3   Mining: Between a Rock and a Hard Place.. 19
4   What Are Our Mineral Priorities?... 25
5   EPA Gives Placer Miners the Shaft 29
6   Grassroots U.S. Mineral Exploration Declines Sharply........................ 35
7   Miners See Light at the End of the Tunnel.. 39
8   U.S. Mineral Exploration Is Being Decimated 43
9   Congress Is Killing America's Economy 47
10 Clinton Launches Attack on American West 51

PART II: MINING AND THE U.S. FOREST SERVICE
11 USFS Mineral Examinations Violate Claimant's Rights 57
12 USFS Responds to Author's Article on Claimant's Rights Issue 63
13 Author's Response to USFS Letters on Claimant's Rights Issue............ 69
14 USFS Proposal Would Bypass Mining Laws... 73

PART III: THE SMALL MINER AS SACRIFICIAL LAMB
15 Nevada Governor Opposes $100 Mining Claim "Holding Fee" 81
16 $100 Mining Claim Fee Proposal Defeated 83
17 OMB Resurrects $100 Mining Claim "Holding Fee"............................ 85
18 $100 Claim "Holding Fee" Deleted from H.R.2686 89
19 $100 Mining Claim "Fee" in 1992 Federal Budget................................. 93
20 U.S. Government Planning to Sacrifice Small-scale Miners 101
21 Small-scale Miners Get the Shaft... 107
22 Small-scale Miners Get Hit with $100 Fee and Bonding...................... 111
23 Locating and Recording New Mining Claims....................................... 117
24 New Mining Claim Fee Policy Implemented 121

PART IV: NEVADA MINING'S FIGHT TO SURVIVE AND THRIVE
25 Nevada's Gold Mining Industry.. 125
26 Nevada's Proposed Gold "Fee" Opposed 129
27 Nevada's Mine Tax Issue Still Undecided.. 135
28 Nevada Mining Tax Bill Signed into Law... 139

29	Anti-mining Forces Focus on the Comstock Lode	143
30	Nevada Politicians Target Mining for Tax Increase	147
31	Nevada Mine Tax Battle Grows Stronger	151
32	Nevada Mine Tax Increase Approved by Voters	159
33	Nevada Legislature Seeks More Mine Tax Revenues	163
34	Nevada Mine Tax Issue Finally "Clarified"	167

PART V: METALS AND MINERALS MINING IN THE SILVER STATE

35	Nevada Enacts Mine Reclamation Law	171
36	Nevada Legislative Actions on Mining Issues in 1989	175
37	Governor Honors Nevada Mining Industry	179
38	Small-scale Mining in Nevada	181
39	Nevada Miners and NDEP Working on Regulatory Changes	185
40	Nevada Will Fight Unreasonable & Unenforceable EPA Standards	189
41	Nevada's Mining Industry	193

Afterword	197
About the Articles' Author	199
Dave W. Parkhurst Mining Writing Collection (Appendix A)	201
Year of Publication in the *CMJ* (Appendix B)	203
References	205
Index	207
Editor Contact Information	219

Foreword

Editor's Note: The Foreword to the Dave W. Parkhurst Mining Writing Collection that was provided in Volume 1 is duplicated here in Volume 4 of the collection because of its applicability to the contents of this volume in particular.

In 1987 I was appointed by then-Governor Richard Bryan to the post of Executive Director of the Nevada Department of Minerals. The Department (now Division of Minerals) and the Nevada Commission on Mineral Resources is a state agency whose mission is to "promote the responsible development of mineral resources in Nevada." Shortly after I started with the State, I was attending a Nevada Miners and Prospectors Association lunch meeting at Grandma Hattie's restaurant in Carson City. It was there that I had the good fortune to meet a man who, I quickly learned, was one of the most rational and effective voices for the small-scale miners and prospectors in Nevada and the west, Dave Parkhurst. Dave became my friend and we worked together as advocates for responsible mineral development many times on many issues until his untimely passing in 1993. Dave's work and steadfast reporting through his articles compiled by Sue Parkhurst in *Fighting the Good Fight* reflect much of that work.

The late 1980s and early '90s was a time of rapid growth in Nevada's mining industry with the discovery, development and expansion of major gold mines in the northern and central part of the state. Large multinational mining companies were doing much of this development, but even with these well-funded industry players, the small-scale miner and prospector played the role that they always have throughout time. As ever, in many cases, original discoveries of mineral deposits were made and mining rights obtained by much smaller entities, and then later sold to the larger companies.

About 86 percent Nevada's surface was then and is now largely managed by the federal government through its US Bureau of Land Management and US Forest Service. This fact, together with the growth of major mining activities in Nevada and the west in general, more than ever attracted the attention of legislators, regulators and the public. Unfortunately, many individuals from these groups would prefer that mining with its impacts on the environment not take place. This is despite the fact that all people, regardless of where they live or how they live need the products made from a variety of minerals that have to be mined from the earth. When you think about it, you'll find the phrase "If it isn't grown, it has to be mined" is true.

By 1990, the stage was set for many changes to the federal and state laws, regulations and policies that governed mining and mineral exploration activities on public land, later moving to private land. Unless there were strong mining advocates with clear and concise positions effectively brought to the state legislatures and the US Congress, mining for hard rock minerals in the west could have been irreparably damaged. The large mining companies had the ability and wherewithal to affect this process to reasonably manage the impact on their businesses, but without people like Dave Parkhurst, it was easy to see the small-scale miner and prospector as an endangered species. While still very challenging, that didn't happen. People still prospect, stake mining claims and make important contributions to keeping the supply of mined products flowing.

Today, while always a challenge in terms of public debate, mining on both public and private land is proceeding in a safe and environmentally responsible manner while having significant economic impacts. This is thanks, in part, to individuals like Dave Parkhurst.

RUSS FIELDS
Reno, Nevada

Russ Fields worked in the mining industry and was executive director of the Nevada Department of Minerals, president of the Nevada Mining Association, and, most recently, director of the Mackay School of Earth Sciences and Engineering (formerly Mackay School of Mines), University of Nevada, Reno. He is now retired.

Editor's Preface

Editor's Note: The prefaces in Volumes 1 to 3 of the Dave W. Parkhurst Mining Writing Collection covered at length how and why this compilation of Dave's written work came to be published. What follows is a summary of the origin and purpose of the collection.

Dave Parkhurst spent much of his childhood in northern Nevada where he explored and prospected in the desert with his grandfather. At age 41 he left behind his previous line of work (in telecommunications) and in early 1980 formed a partnership with two friends in a mining venture which, while it didn't "pan out," indirectly led to his beginning a new, and final, career. In the depths of the recession that began in 1981, Dave started writing a weekly column on mining and prospecting for the *Mountain Messenger* newspaper in northern California's Mother Lode country. By early 1982 he had become a regular writer for the *California Mining Journal* (*CMJ*, now the *ICMJ*).

For the next 12 years, from 1981 to 1993, Dave was engaged in the field of mining and prospecting as a full-time writer and consultant, and eventually as a lobbyist, too (in behalf of Nevada's small miners and prospectors). During this period he also devoted whatever time and resources he could manage to the acquisition and development of his own mining properties. On a September day in 1993 he was preparing for his upcoming trip to the mine that he was a partner in when he died suddenly from a massive heart attack.

The Dave W. Parkhurst Mining Writing Collection is my attempt as Dave's widow to honor my late husband by preserving his written work, in book form, as an important part of his legacy. At the time of his death, Dave was a recognized expert in small-scale mining and prospecting. He was knowledgeable in many facets of metals and minerals mining, particularly in the area of exploration, and had mentioned on at least one occasion during our 13 years together the idea of writing a book. While this compilation of articles that he wrote for the *CMJ* is not a substitute for the book that he never got to write, I think he would have been pleased with it. The collection consists of about 150 articles distributed among four volumes; the remaining articles and other writings can be found in the archives of the website being developed as a repository for Dave's written work.

Editor's Acknowledgements

My thanks to all those who helped in some way to make the Dave W. Parkhurst Mining Writing Collection possible. The specific individuals, whose contributions pertain to the entire collection rather than its individual volumes, are acknowledged in Volume 1, and the particulars of their contributions to this project can be found there.

FIGHTING THE GOOD FIGHT

MINING'S BATTLE FOR SURVIVAL IN THE AMERICAN WEST

VOLUME 4

Dave W. Parkhurst
MINING WRITING COLLECTION

Introduction

Volume 4 of the Dave W. Parkhurst Mining Writing Collection represents the activist side of the work that Dave was engaged in as a writer and consultant in the mining field, advocate for mining, and lobbyist at the Nevada Legislature in behalf of his fellow small miners and prospectors.

Fighting the Good Fight: Mining's Fight for Survival in the American West, the title of this volume, aptly describes the mining industry's fight for survival against the forces arrayed against it in the 1980s and '90s. Dave Parkhurst played an active, consequential role in that struggle.

Freedom and individual liberty were the ultimate causes for which Dave fought and which he valued above all else. Freedom is experienced and expressed in many ways, including the freedom to make a living in a legal enterprise that provides great value to society—that is, mining—without being constantly targeted for extinction by hostile and irrational forces in the name of environmental protection and utopian ideals. It also includes the right of American citizens to access and utilize resources on public lands, as allowed by law and under reasonable restrictions. These freedoms have been in jeopardy even in America, Land of the Free, for many years, and Dave was fully engaged in the fight to preserve them.

This final volume of the four-book collection is divided into five parts, each with a particular variation on the theme of "fighting the good fight." The articles in Part I consist of commentary, essays and reports on battles faced by the mining industry and other natural resource industries. Among these battles were many that concerned environmental protectionism and wilderness withdrawals. As an outdoorsman himself, Dave understood the necessity and desirability of protecting the environment and preserving a portion of the nation's lands as wilderness; he supported *reasonable* environmental protection and conservation efforts.

Part II focuses on an episode involving a miner's interaction with the U.S. Forest Service and the sometimes contentious relationship between the mining community and the government agencies whose regulatory overkill adversely affects miners and prospectors. The article, titled "USFS Mineral Examinations Violate Claimant's Rights," inspired letters to the editor from Forest Service personnel that defended the agency and its efforts to work with small miners and prospectors.

Part III consists of a series of reports, as well as commentary, on the federal government's ultimately successful imposition of a $100 mining claim holding fee on miners and prospectors. In its final form, the legislation enacting this fee threatened to take a heavy toll on minerals exploration in the U.S., in no small part due to its effects on small-scale miners and prospectors. This sequence of articles illustrates in condensed form the nature of the battles and challenges confronted by America's miners and others in the natural resources industries.

In Part IV, "Nevada Mining's Fight to Survive and Thrive," the articles chronicle the battles over tax-and-control issues that had to be fought by the mining industry in the mid-'80s to early '90s. These battles were fought in the Governor's Office, the Nevada Legislature, and the news media.

The last group of articles, in Part V, concerns various topics related to mining in the Silver State, including the determination of the state to fight unreasonable and unenforceable EPA standards and the honoring of Nevada's mining industry by a former adversary, Governor Bob Miller.

The reader is reminded here that the contents of Dave's articles reflect both fact (objective reality) and the author's opinions on the issues, based on his observations and personal views. Also, keep in mind that the chronicled events took place more than two decades ago—some of them, three decades ago—and the articles, written in the 1980s and early 1990s, may no longer accurately reflect the state of the nation's mining industry and the conditions under which today's small-scale miners and prospectors operate.

In any event, the articles in this fourth and final volume of the Dave W. Parkhurst Mining Writing Collection tell a compelling story, in a continuation of the theme of Volume 3, of the fight for survival of metals and minerals mining in the American West by its advocates, including Dave Parkhurst, in the years 1982-1993.

PART I

The U.S. Government vs. Mining

"Although government has consistently shown that it is totally incapable of managing itself, it persists in the belief that it can better manage all businesses and each individual's personal life."

1
Economic Policies and the Mining Industry

Author's Note: I recently wrote several articles concerning the various factors affecting the mining industry and/or metals and minerals prices (i.e., "Precious Metals Outlook for 1985," Dec. 1984 *CMJ*, and "The Next Mining boom," Jan. 1985 CMJ). The projections made in these articles were predicated on some necessary changes in political and economic policies that are obviously required for a healthy U.S. economy. Unfortunately, it appears that these changes in policy are either being ignored or delayed. It seems appropriate, therefore, to call attention to several of the political and economic policies currently being pursued that might produce disastrous effects upon the minerals sector.

CONTRARY TO THE BELIEF that the high U.S. federal deficit has caused an increase in interest rates and an inflated value of the U.S. dollar, interest rates over the past few years have been purposely predetermined by financial policymakers (the Federal Reserve Board and banks). A "tight money" policy has been placed into effect with the purported objective of controlling inflation, and this action also produced a shortage of dollars in the marketplace. Neither interest rates nor the value of the dollar have been allowed to be determined by the normal supply and demand forces operating in a free market economy. With this type of monetary control, most capital investment has been drawn into "paper transactions" and very little has gone into the production of hard goods or industrial modernization and expansion. This, in turn, means there has been a sharply reduced demand for raw materials used in industry—including the minerals and metals produced by the mining industry.

It is also beginning to look like the U.S. federal deficit will be surpassed by the foreign trade deficit. The imbalance in foreign trade is progressively becoming worse, and most economists predict that this will continue to be the case over the next few years. Record foreign trade deficits also mean that the U.S. is effectively exporting thousands of jobs and billions of dollars needed to bolster the U.S. economy. In addition, this country is rapidly becoming oriented towards the production of "services" rather than the production of hard goods. And most of the capital available for the purchase of raw materials is being spent in the foreign commodities markets.

This brings us to the political and economic policies responsible for the expanding foreign trade deficit. Over the past decade, billions of dollars have been spent to industrialize the economies of many small foreign countries. This policy was initiated by the U.S. government and the International Monetary Fund in an attempt to raise the standard of living and bolster the economies in the so-called Third World nations. A net result of these policies has been the strengthening of foreign industrial capacity while gradually weakening this nation's industrial strength. In effect, these policies created our current competition in commodities production and sales, with no provisions being made to ensure that industries in this country could remain competitive on the world markets.

Which brings us to the tremendous U.S. federal deficit. Obviously, if we are spending billions of dollars on foreign commodities, we cannot spend those same billions within this country. The expenditure of massive amounts of capital on hard goods generates a large amount of tax revenues for the government by expanding the tax base and creating a large number of jobs. This means that the federal deficit would be reduced by any reduction in the foreign trade deficit. But, our analysts tell us, this is not to be the case—they are predicting that the foreign trade deficit will *increase*. And this, in turn, is caused by the fact that our industries cannot become competitive on the world market—*primarily because of tight money and high interest rates*. As mentioned, these same monetary policies and rates charged for interest are set by our own financial institutions.

And now, our economic problems are about to be compounded. Over the past year, the media and most other informational sources have been painting a biased picture of the overall health of the U.S. economy. This was most certainly done to give the consumer a false sense of security and encourage consumer spending so as to buy the United States out of the economic recession. According to recent figures, however, the consumer is beginning to realize that the picture is not quite as rosy as it first appeared to be and is starting to salt away more money in savings and investments as a hedge against future economic problems. The Federal Reserve Board, largely responsible for setting interest rates and controlling the supply of money, has noted the increase in capital available for spending, and has decided to continue restrictive financial policies in an attempt to get the consumers to spend their money (again with the purported objective of controlling inflation). In this case, however, the consumer is not likely to be misled again—and will probably continue to save and invest as much capital as possible.

If the Federal Reserve Board does not realize that the economic situation has changed, and persists in holding to its narrow range of monetary growth allowed for a so-called healthy economy, it will continue to persist with tight money policies and high interest rates. This action would effectively choke off any possibility of a continuing economic recovery, would ensure that the foreign trade deficit increased, and would negate any attempts at bringing the U.S. federal deficit under control.

In order to achieve sustained economic growth, it is necessary that U.S. business be encouraged and allowed to expand and modernize at a fairly rapid rate. To accomplish this objective, it is also necessary that a sufficient amount of capital is available for expansion at reasonable interest rates. This, unfortunately, is apparently not to be realized in the near future. Several economists have recently published figures that show the U.S. business sector plans to expend about half the capital invested last year in industrial expansion and modernization during 1985. The main reasons cited for reduced business investment were high interest rates and expectations of lessened demand for domestically produced goods in the U.S. The figures also show that most of the expansion capital is to be spent on machinery and finished goods—almost all of which are to be purchased from foreign countries. These conditions will serve as a depressant upon metal and mineral prices and will also further reduce overall demand for mineral commodities and raw materials on the world markets.

Under normal conditions it might be expected that the billions of U.S. dollars flowing into foreign countries would be expended in the marketplace and eventually provide additional demand for American goods and services. This is not to be, however, as most of the nations receiving foreign trade dollars are financially strapped by their huge debts to the IMF and some U.S. banks. Practically all of this capital goes towards paying principal and interest on these debts plus maintaining economic stability within the foreign countries involved. The net result of this process is another paper transaction which returns much of the money to the same financial institutions that are maintaining tight money policies and high interest rates—instead of injecting the capital into the economy where it might alleviate the business problems in this country. And so the cycle will continue, unless some immediate and realistic changes are made.

It is interesting to note that maximum attention is still being focused upon the U.S. federal deficit as being responsible for continued high interest rates and an overvalued dollar. Yet the money supply and interest rates are being directly controlled by the Federal Reserve Board, which is an institution devoid of any federal control. And, when questioned about their strict monetary policies, financial institutions continually throw up the specter of "renewed inflationary pressures." How is it that the financial policymakers can say that interest rates are determined by the availability of investment capital, when they not only control the money supply but also decide upon what the interest rates are to be in advance?

Another interesting aspect of the economic policy spiral is the fact that encouraging purchases of foreign commodities also ensures that payments will be made on Third World debts. This means that the same policies that are damaging the U.S. business environment also act to ensure against a potential default on debts owed by foreign countries, and it also tends to strengthen foreign industry at the expense of U.S. industrial strength. In addition, it encourages foreign overproduction of metals and minerals that must be sold on an already depressed market, further compounding the problems currently being faced by the mining industry.

The continued pursuit of the current disastrous economic policies can only lead to either another recession or a full-scale depression in the United States. Unless or until our business and industry is allowed to recover and expand, our economic health has only one way to go—downward. The mining industry is especially vulnerable at this time, because in order for mining to recover it is first necessary for industry to recover. In other words, a demand must exist for mineral commodities before they will be produced. And the normal time delay between an increase in demand and an increase in minerals production is usually measured in years.

Under the present circumstances, even a "quick fix" would take a considerable amount of time to work its way down to the producers of raw materials for industry. So even if less restrictive monetary policies were adopted today, it would be some time before there would be any noticeable impact upon the mining industry. And it does not look likely that any significant changes will be made by our financial policymakers in the near future.

There is some action we can all take, however, that might help to alleviate the current economic situation. We can write, or telephone, our congressmen and representatives and encourage them to pass legislation granting governmental control over the Federal Reserve Board, and also to authorize the president to exercise the line-item veto on bills passed by the Congress. This is the only free-market-economy country in the world that does not have some form of governmental control over its own internal financial and economic policies.

2
The Great Economic Sacrifice

Author's Note: The following article is by no means complete in every detail. It is intended to illustrate the overall effect of current economic policies upon the mining and manufacturing sectors, as well as the rest of the U.S. economy, and to update and clarify information contained in "Economic Policies and the Mining Industry" (Feb. 1985 *CMJ*).

ECONOMIC ANALYSTS AND FEDERAL officials are finally beginning to tell us what we have known for some time: the mining and manufacturing industries are in trouble. And this is not because these industries have failed in their efforts to remain competitive on the world market but is almost solely a result of current U.S. economic policies. As evidenced by rock-bottom commodity prices and escalating foreign trade deficits, the primary problem is the "strong" dollar. Most of the hard goods, natural resource, farming, and high technology industries are declining for the same reason. The impact upon mining and manufacturing has been particularly severe, as their traditional markets are rapidly being cornered by foreign producers.

Erroneous information (or "disinformation") has led the majority of the American public to believe that an overvalued dollar is a good thing. It has been touted as the means whereby the U.S. experienced an economic recovery while holding inflation in check. Cheaper foreign goods, and sharply reduced demand for U.S.-made products on foreign and domestic markets, *have* acted to hold down price increases for most commodities. But what price have we paid for this temporary reduction in inflation—over both the short and long term?

In order to obtain a clear picture of what has been happening to the U.S. economy in recent years, we should first take a look at the original problems and the actions taken to correct them. Rampant inflation, which resulted primarily from high oil prices and unrealistic demands by the general public and special interest groups, was countered by increased deficit spending for government programs. In a cause-and-effect reaction, both of these factors fed upon each other in an uncontrollable upward escalation. When this inflationary economic spiral became global in scale, it collapsed upon itself and caused a world-wide recession in the early 1980s.

Simply put, demands for higher wages, more benefits, increased aid and welfare programs, higher social security payments, and so forth, plus increased costs for environmental concerns, raw materials, oil and energy, taxes, transportation, etc., led to dramatic price increases. This, in turn, led to increased deficit spending by government to procure its goods and services, as well as to meet escalating demands for more costly social, welfare, aid and work programs. Many other factors were also involved, particularly wasteful and inefficient fiscal policies.

In their ineffable wisdom, American political and financial wizards hit upon the ultimate defense against inflation: creation of the new Super Dollar. The resultant financial and economic policies achieved three immediate results: (1) financing of the continuing huge federal deficit with foreign funds; (2) a sharp reduction in the competitiveness of U.S. industry on foreign markets—to the point that this sector of the economy is in danger of collapse; and (3) an increase in the total debt load of economically strapped Third World countries. This last item ensured the continuing inward flow of cheap commodities for U.S. consumers and created massive interest payments to U.S. and world financial institutions which helped to offset their losses from bad loans and mismanagement.

These policies also froze prices at the prevailing high levels while drastically reducing prices for raw materials—thereby tending to hold inflation in check by artificially increasing the potential profit margin for finished products. Most importantly, they provided the consumer with an illusion of a sustainable, anti-inflationary economic recovery.

The Super Dollar was created by intentionally controlling interest rates and the basic supply of money in circulation—*not*, as is currently cited, by the pressure exerted by the federal deficit on the world's financial markets. This has been demonstrated by the fact that interest rates have gone *downward* over the past year while, during the same period, both the federal and foreign trade deficits have *increased dramatically*. Interest rates, however, are still unreasonably high, and they are being maintained at that level to support the inflated U.S. dollar and ensure the inflow of foreign investment capital.

This monetary policy places the U.S. economy in an unstable and unrealistic position. We now rely on *foreign* capital to finance the federal deficit and keep the faltering economy going; we are relying on increasing amounts of *foreign* credit to finance our increasing imports of *foreign* products and raw materials; and we are beginning to rely upon *foreign* investments to bolster U.S. financial institutions and industries. In other words, we are accumulating a massive foreign debt in order to support our overall economy today, and in the future. Just who do we think is going to pay the principal and interest on *these* debts?

A few excerpts from a July 23, 1985, Associated Press news release illustrate current views on some of the factors involved:

> (Federal Reserve Board Chairman Paul) Volcker made it clear that the central bank was not interested in fostering further declines in the dollar through (changes in) its monetary policies as long as the congressional budget impasse remains unsolved.
>
> "The hard fact remains that so long as we run massive budgetary deficits, we will remain dependent on unprecedented capital inflows to help finance, directly or indirectly, that deficit," Volcker said.
>
> (Commerce Secretary Malcolm) Baldrige, however, said that for U.S. industry to be competitive again against foreign manufacturers, the dollar needed to fall much farther than it had recently.
>
> "For American industry to be completely competitive again, we would like to see another 20 to 25 percent decline," Baldrige said.

These statements acknowledge both the adverse impact of an inflated U.S. dollar and the accumulation of foreign debt. It is politically expedient, however, to continue the same detrimental monetary policies in order to compensate for the inability of Congress to reduce deficit spending. In other words, our economic security is being undermined by the weakening of U.S. basic industries and the accumulation of massive foreign debts—with the full knowledge of those who are in control of our monetary policies. This amounts to the sacrifice of certain elements of the U.S. economy, particularly mining and manufacturing, in order to protect those elements having "higher priority."

Now, let's look at the nuts and bolts of this sacrificial policy. It has been estimated that well over a million high-paying jobs in the mining and manufacturing industries have been virtually exported to foreign countries. What provisions were made to compensate the displaced workers in these sectors of the U.S. economy? They were supposed to be employed in "new" jobs created within the services sector, which often pay less than half of the amount these workers were receiving in their previous jobs. Many of these people were also considered as unemployable without some retraining for their new occupations—largely because of the many years of experience and specialization acquired in their previous trades. Is it fair that these people should be required to give up their areas of expertise and job security, in exchange for a lower standard of living and jobs that many of them do not want or like?

Keep in mind that this "transfer" of highly qualified personnel to service-oriented jobs is continuing at the present time, with an estimated 220,000 manufacturing jobs lost between March and July of this year. This is in addition to the employment lost in mining, farming and other sectors being decimated by the present economic policies. Many of these people are only employed part-time, yet the government statistics reflect them as being in the same category as full-time workers.

The federal government has even initiated special training programs for certain workers, especially miners, because officials do not expect that these industries will be able to fully recover their previous strength. And this is accepted as being worth the "benefits" of an inflated dollar?

There are, admittedly, several other economic factors involved in the decline of U.S. mining and manufacturing, particularly the previous U.S. administrative and financial policies designed to assist in the development of viable mining and manufacturing industries in many of the underdeveloped nations. This disastrous experiment created many of the same industries that are now providing unfair competition with the remnants of this nation's mining and manufacturing sectors, as well as saddling several poor countries with huge debts.

Many of these foreign industries are subsidized by their governments, operate with cheap labor, and sell their products at less than the cost of production to obtain much-needed foreign exchange capital—most of which goes towards interest payments on their debts. When considering all of the factors involved, however, U.S. industry would still be fairly competitive on the world markets—without the trade barrier imposed by an artificially inflated U.S. dollar.

The foreign trade deficit is accumulating massive foreign debt at an accelerated pace. Our struggling economy is, in effect, mortgaging its future, as the United States heads towards the distinction of being the most debt-ridden nation on the planet. At the same time, our present and future ability to repay these debts is being rapidly destroyed. While most foreign nations are encouraging and expanding their basic industries, this country's financial and economic policies are continuing to undermine our industrial capacity and economic stability. How can we justify this sacrifice?

The critical nature of these problems has been apparent to business leaders for some time, but their complaints to our political and financial leaders have apparently been ignored. Our basic industries have been told they must be "more competitive" on the world markets, while our monetary policies have continued to make that objective impossible to attain.

Under present conditions, this country's mining and manufacturing sectors would have to be almost *twice* as cost-effective as foreign producers in order to overcome the foreign exchange imbalance created by the inflated U.S. dollar. Recognizing this insurmountable obstacle, many U.S. businesses are now actively seeking the imposition of trade barriers and "protectionist" policies. Such actions could not alleviate the trade problems for any appreciable period of time, and they would undoubtedly result in foreign retaliation of the same kind. It would also ignore the underlying cause of the trade imbalance.

This brings us to the crux of the entire issue. Will this country allow its mining and manufacturing industries to be sacrificed for the temporary illusion of economic health and stability? Will we allow the continued pursuit of disastrous policies that undermine our economic strength and national security? Or will we attack the root causes of our problems: federal deficit-spending and the inflated U.S. dollar?

3
Mining: Between a Rock and a Hard Place

DESPITE ITS CURRENT DIFFICULTIES, the North American mining industry remains fairly optimistic about the long-term potential of mineral resource development. This is a rather surprising attitude in an industry that has been systematically decimated through the combined effects of faulty fiscal and monetary policies, overwhelming regulatory controls, and political and economic policies designed to encourage worldwide overproduction of cheap mineral commodities.

If this seems to be a biased viewpoint, consider that our manufacturing industries, farming, ranching, many financial institutions, and other businesses are in dire straits for the same basic reasons, and the ripple effect has begun to seriously impact our high-technology and service industries. For many industries and the overall economy, the worst is probably yet to come, because many political and financial leaders are either unwilling or unable to seriously address the underlying causes of these problems.

The North American mining industry has been undergoing a steady decline for the past six years. Why did this massive downturn occur in an industry that occupied a preeminent position in the world's business community just a few years ago? The answers lie in the general statements made above, but the full explanation is more complex.

The initial downward trend was started by the worldwide recession of 1981-82, which precipitated an effect most miners know as the cyclical nature of mining. Simply put, this term covers the normal boom-and-bust periods inherent in the minerals-producing sector. Economic factors dictate that the industry has boom periods when mineral prices rise because the demand for mineral products exceeds the supply, and bust periods when the situation is reversed. Thus, when industrial activity and consumption patterns dropped sharply, so did minerals demand—and the mining industry itself.

Under normal conditions, when the industrialized economies began to recover from a recession, the resultant increase in minerals demand would have produced a subsequent rise in mining activity. Conditions were *not* normal, however, and the minerals sector found itself faced with an even more drastic decline: more mine closures, layoffs, weaker mineral prices, production cutbacks and increased capital losses. The basic economic factors impacting the mining industry had undergone a radical change.

A number of negative forces contributed to the almost 50 percent reduction in U.S. minerals production capabilities, but two of them could have done it alone: foreign minerals competition and the overvalued U.S. dollar. The combination of cheaper foreign minerals and a strong dollar became a two-edged sword: as the foreign currency values fell against the dollar the mineral commodities became even less expensive, and as the price of mineral materials fell the profit margin for consuming businesses rose—thereby bolstering the U.S. economy, propping up the dollar, and maintaining higher consumer prices.

A drop in the cost of raw materials usually produces a consequent reduction in wholesale and retail prices, making finished products cheaper (they cost less to produce) and creating more consumption and demand. The increase in consumption and demand for goods translates into an increased demand for minerals and raw materials, thereby causing increases in the prices paid for minerals. Not this time, however.

The rampant inflation that preceded the recession had produced an immense rise in the overall price structure, so retailers maintained the prevailing higher price levels and utilized the lower production costs to increase profit margins. This action gave the illusion that the economy was rebounding (through higher profits and increased employment in selected businesses, mostly service-oriented) and ensured the continued decline of basic industries, farming and mining. Why? Because the prices for raw materials remained artificially depressed.

Foreign imports then began to flood the U.S. markets, for two basic reasons: (1) commodity prices in foreign currencies increased dramatically as the dollar rose in value (foreign producers received more of their currency per commodity unit), and (2) U.S. and World Bank financing had created more minerals producers by making huge development loans to the Third World countries. In other words, we had created the competition and then gave them the incentive to increase production and exports—thereby forging *another* two-edged sword.

As the U.S. dollar continued to rise (due to higher interest rates), basic mineral commodity prices continued to fall. This happened because no significant increase in minerals demand was being created by industrialized nations due to their maintaining higher prices for finished goods (i.e., discouraging increased consumption and resultant demand), and the foreign minerals producers were flooding the world markets with an overabundance of cheap mineral supplies and other goods. Here's where the second two-edged sword entered the picture.

Under the usual economic circumstances, as the world mineral prices fell below the cost of production, foreign producers would have reduced their output. Not so this time, however, because they were now saddled with so much debt that they were forced to continue overproduction—just to make payments on their debts and keep their economies going. In addition, many state-owned or state-controlled mineral producers in the developing countries and centrally planned (communist) economies often continue to maintain uneconomic production in order to sustain export earnings, provide employment, and generate required tax earnings—regardless of the inevitable consequences to the world's economy. Both types of governments now found themselves with an unsolvable dilemma: as they continued to swamp the world market with an oversupply of minerals, prices were pushed even lower, which forced these countries to make additional increases in production and exports to make up for the reduction in earnings—thereby further depressing the world's commodity markets to record lows.

All of these factors eventually produced record high levels of cheap foreign imports into the U.S. (further decimating our mining, manufacturing and farming industries) and record low U.S. exports to foreign nations (which also knocked the wind out of our resource industries).

At this point, with U.S. industry on the ropes, our financial wizards finally decided to drop interest rates and the value of the U.S. dollar on foreign currency exchanges, predicting that this action alone would reduce the foreign trade deficit and save our ailing industries. But this proved to be too little, too late, and to no avail.

Why didn't this policy work? For two reasons again: (1) foreign producers weren't about to give up their newly acquired share of the world commodity markets, so they dropped prices whenever and wherever necessary, and (2) the "devaluation" of the U.S. dollar was only targeted against the currencies of our high-tech and service trading competition—none of whom are major minerals or raw materials producers or exporters.

Now, back to the underdeveloped and least industrialized nations: While the dollar fell dramatically against the currencies of major industrialized countries, it stayed about the same (or increased) in value relative to the currencies of most major minerals and raw materials exporting nations. So, these countries are still continuing to overproduce and glut the commodity markets. In their currencies, most mineral commodity prices have increased slightly as a result of currency disparities in some of the foreign industrialized nations' exchange rates—which were, in turn, brought on by the efforts to drop the dollar's value relative to those currencies (Japan, West Germany, France, Britain, etc.). These developed nations are also major importers of minerals and raw materials. In addition, Third World countries have no incentive to reduce production and exports, because of their tremendous debt burdens and faltering economies.

So, where does this leave mineral commodity prices? Rock bottom. And where does it leave our base-metal mining, smelting and refining industries? Between a rock and a hard place.

All is not gloom and doom, however, as some segments of the mining industry were able to retrench, cutting operating costs and improving production efficiencies. Precious metals mines fared better than most, since this percentage drop in gold and silver prices (in U.S. dollars) was not as severe as the plunge in most base-metal prices. Primary silver producers were impacted more than gold producers, as foreign producers continued to dump an oversupply of silver on the market. Many of the more marginal precious metals mines were forced into temporary closures, as were some of the major silver producers.

The mining industry's immediate future still has one bright spot: precious metals production. This segment of the industry is expected to be the first to climb out of the current slump. In addition, several major base-metal producers (particularly copper) are now holding their own and are beginning to strengthen. Overall, the North American mining industry has shown a remarkable amount of resiliency and underlying strength in the face of almost insurmountable odds against its survival. And it has done it alone.

The challenge to the mining industry today is to remain competitive in the world markets while still being subjected to severe environmental constraints and overregulation, as well as facing increased foreign minerals competition. Since most foreign producers are not burdened with any environmental concerns, our minerals industries will have to become much more efficient and cost-effective than their foreign counterparts in order to survive. Although it will be a sometimes painful (and often one-sided) battle, the North American mining industry can, and will, meet the challenge.

4
What Are Our Mineral Priorities?
(1988)

MINERALS AND RAW MATERIALS represent both the real wealth and the basic ingredients of our modern civilization, and their influence on international politics is immense. Further, a direct relationship exists between our security and well-being as an integral society and the effectiveness with which we utilize our available mineral resources.

Legislators in the United States have been aware for some time that the country is far too dependent upon foreign sources for many of the raw mineral commodities and refined metals necessary to maintain the economy and national security. This realization of the critical nature of the matter has resulted in several policy changes and legislative actions that were supposedly designed to correct the problem, by making the U.S. more self-reliant in minerals production—particularly in the area of strategic and critical minerals and materials. Why, then, is the overall trend towards an *increasing* reliance upon foreign sources for supplies of these mineral commodities allowed to continue?

Instead of encouraging growth in the domestic minerals industry, the exact reverse is being accomplished. Through excessively strict environmental regulations and standards, the wholesale closing off of increasing amounts of the public domain lands to mineral exploration and development, the proliferation of legal and administrative overregulation, the constant addition of new taxes and required regulatory "fees," the onerous burden of rampant litigation, and the ongoing subsidies being provided by the U.S. taxpayers to promote foreign minerals development, the U.S. minerals industry is being squeezed out of the international markets. These actions are not only detrimental to the national economy, they also undermine our national security. Consider the following:

Between 1938 and 1968, the United States used more mineral commodities and fuels than the entire world used in all of its previous history. In the last 20 years, this use has more than doubled. There are now at least 64 minerals and metals for which the U.S. is a net importer from foreign sources, and 29 of these commodity imports are among the 93 strategic and critical minerals included in the National Defense Stockpile.

The nation is now 100 percent import-reliant for supplies of ten metals and minerals (arsenic, cesium, columbium, graphite, manganese, sheet mica, rubidium, strontium, thallium and yttrium), and it is well over 80 percent import-reliant for another ten metal and mineral commodities (gem stones, bauxite and alumina, bismuth, gallium, tantalum, diamond, fluorspar, the platinum group metals, cobalt and tungsten). This list could be expanded considerably if the remaining metals and minerals upon which the United States is largely import-reliant were included. More importantly, several of the foreign countries that supply these commodities are either politically or economically unstable (or unreliable). What would happen if the supplies of these materials were cut off?

As the U.S. population grows and our standard of living increases, the demand for metals and minerals will also continue to increase. As the world population grows, and the presently underdeveloped countries raise their standards of living to match that of the U.S., their mineral demands will also increase. In 1900, the world population stood at about 1.75 billion. Today, the world population is 5 billion, and it is expected to double to a projected 10 billion people in the not-too-distant future. It is difficult to imagine how this tremendous increase in population will impact the supply and demand picture for metals and minerals, but it is obvious that new sources of supply must be discovered and developed. Considered in the light of its current minerals policies, is the U.S. prepared to meet this challenge?

Intense mining activity in the U.S. during the past 200 years has resulted in a situation where most ores being mined in this country are of a substantially lower grade than those being mined in many foreign countries. In addition, as the more accessible ore bodies are depleted it then becomes necessary to mine more of the metals at a greater depth. This problem is much more immediate when the strategic and critical minerals are considered, since most of the U.S. ore deposits are already lower in average grade, fairly localized, and occur at greater depths, which makes them much more difficult and costly to mine. Some of the major problems involved in mining deeper ore deposits include rock bursting, squeezing ground conditions and much higher temperatures. As a result, new mining methods, ground control techniques, materials handling systems and environmental controls will have to be developed in order to exploit these deeper deposits and decrease the nation's dependence upon mineral imports.

It is obvious that the U.S. should maintain a dependable supply of mineral resources that is sufficient to meet its economic, societal and defense requirements. However, the nation's minerals industry cannot meet that responsibility by itself, mainly because most of the supply and demand factors are substantially influenced by national and international issues and policies that are the responsibility of the government. In addition, the factors noted above are tending to restrict, rather than promote, development of a healthy domestic minerals industry.

The nation is continuing to remain vulnerable to potential disruptions in its supplies of metals and minerals, especially in regard to the critical and strategic minerals. If this situation is allowed to continue, or worsen, then our national security and economic stability may be placed in further jeopardy. The current situation is still intolerable, and it must be corrected before it becomes even more critical.

Unless this trend is reversed, we also risk the possibility of inflicting permanent damage upon the nation's domestic minerals industry. Unfortunately, very few people are aware of the importance of a strong U.S. mining industry in relation to matters affecting our national security and economic well-being.

The time has come when we should ask ourselves: What are our mineral priorities?

5
EPA Gives Placer Miners the Shaft

IN YET ANOTHER ATTEMPT to accomplish legislation by means of administrative legislation, the Environmental Protection Agency has risen to new heights in its efforts to put the nation's placer mines out of business. The agency's new standards for "effluent limitations," purported to fall under provisions of the Clean Water Act and a "consent decree," as specified in 40 CFR Part 440 (frl-3361-7) "Ore Mining and Dressing; Point Source Category; Effluent Limitations Guidelines, Pretreatment Standards, and New Source Performance Standards," would put Mother Nature out of business.

As usual, the EPA has decided to set the standards for allowable Total Suspended Solids (TSS) in discharged waters at a level well below the average levels commonly occurring in the natural environment. In order to justify their authority to regulate common (natural) substances (such as silt, clays and other particulate matter) in the nation's streams and rivers, the agency has conveniently classified them as "conventional pollutants." In other words, the particulate matter contained in normal spring runoff water and above-normal flows following storms is now considered to be "pollutants." So, any time you see muddy or murky waters produced by Mother Nature during periods of high runoff, you can be comforted by the fact that our country's EPA now regards these naturally occurring events as rampant pollution of the environment.

Since natural substances are now polluting the environment, it follows that the EPA must adopt stringent regulations to ensure that placer miners do not add any significant amounts of these pollutants to the nation's streams and rivers. Using the same line of reasoning, the EPA has decided to set standards for discharge waters from placer mines at a level nearly equal to the standards for drinking water. In order to accomplish this purpose, the agency has conducted numerous studies on costly and time-consuming methods by which they believe they have proven that it is indeed possible for most placer mines to meet the agency's standards. In the process, the EPA has formulated a new set of regulations and requirements that would confuse a highly competent attorney. Yet they fully expect the average placer miner to comply with them.

Most importantly, however, is the means whereby the agency attempted to establish the new regulations. No notification was given to the affected parties, no hearings were held for public input, and, in fact, only a few placer miners became aware of the new regulations after they were already in effect. Most placer miners are still unaware of the new requirements, since no attempt was made to contact them. The actual sequence of events is as follows:

The *Final Rule* outlining the EPA regulations was published in the Federal Register, Vol. 33, No. 100, on *May 24, 1988;* the regulations were considered issued for purposes of judicial review on *June 7, 1988;* and, *the regulations became effective on July 7, 1988.* In addition, the EPA stated:

Under section 509(b)(1) of the Clean Water Act, judicial review of this regulation can be made only by filing a petition for review in the United States Court of Appeals within 120 days after the regulation is considered issued for purposes of judicial review [by Nov. 5, 1988].

It was stated further that under section 509(b)(2) of the Clean Water Act, "*the requirements in this regulation may not be challenged later in civil or criminal proceedings brought by (the) EPA to enforce these requirements.*" Additionally: "The record for the final rule will be available for public review not later than *July 28, 1988,* at the EPA Public Information Reference Unit, Room 2904 (rear) (EPA Library)." Note that the record of the final rule for public review *became available 21 days after the effective date of the new regulations.*

Unless each individual placer miner just happened to be an avid reader of the Federal Register, how would it be possible for most miners to become aware of the new regulations—let alone have time to respond to them?

It is quite obvious that the EPA intended to put the regulations and requirements into effect *before any placer miners became aware of the fact that the "final rule" had even been issued.* This is another classic example of the current attempts being made by several governmental agencies to accomplish legislation through the implementation of administrative regulations, but, in this case, the new regulations were not even *proposed.* For all practical purposes, they were *enacted.*

Because I have not yet had sufficient time to fully comprehend and analyze the overall impact of these new EPA regulations on the placer mining industry, I will include several excerpts from comments sent to *CMJ* by concerned placer miners.

Thomas S. Bonn, president of Bonanza Mining Co, Inc., states:

> We feel that if the bureaucrats at the EPA had more concern for the American producer, they would figure out how to effect water quality (standards) without oppressing the producer with unnecessary financial and operational burdens.

In a letter to the EPA, Bonn describes the (substantial) impact of the regulations on his company's placer operations and also states:

> I believe it is futile for the EPA to prescribe a simple recipe that could be effectively applicable to all gold placer mines, because the variance in the geography, geology, soil conditions, etc. is too great. Instead, the local permit-issuing authority should have the latitude to make decisions based on case-by-case applications, according to the existing conditions. The issue should be the final quality of the water discharged into the stream, not how that quality is affected.
>
> To recirculate 100 percent of the wastewater according to EPA standards would require about 25,000 gallons of diesel fuel per year at our gold mining operation. Thus this mine, which the EPA apparently didn't (even) know existed, would require over 5% of the total increased energy requirements of 485,200 gallons of diesel fuel per year that the EPA estimates their proposed BAT technology will require for all gold placer mines in the entire United States.

He also outlined the technical aspects of his company's mining operations. He pointed out the fact that they have been using a flocculant for settling solids from wastewaters on a full-time basis, whereas the EPA flatly states that this is not being utilized by any placer mines on a season-long, commercial basis.

Willard H. Lynch, an engineer, says:

> I hope that all miners understand that it is now necessary for them to obtain an EPA National Pollutant Discharge Elimination System (NPDES) permit, regardless of how little they pan, sluice or dredge. Even the recreational or prospecting miner must obtain a permit before commencing any operation, regardless of the insignificance of the amount of placer they work. You will have to allow 180 days (6 mo.) for the issuance of the permit by the EPA, and the maximum penalty for not having one is $10,000 and six months in a Federal prison. Should you move to another location, you are considered a new operation, even though it is the same owner-operator and the same equipment, and a new permit is required. You must not commence an operation until the permit is issued.

CMJ Editor's Note: Mr. Lynch cites the following wording in the Federal Register as the basis for his conclusion on permit requirements: "Furthermore, the final rule does not establish effluent limitations guidelines and standards for dredges processing less than 50,000 cubic yards of ore per year... [The] EPA emphasizes, however, that *the exclusion of these facilities from the coverage of this regulation does not alter their duty to obtain NPDES permits for their discharges.*" CMJ is seeking further definitive clarification on this.

Lynch also included a copy of a detailed analysis he prepared and submitted to the EPA, in which he outlines the overall impact of the new regulations. The analysis and comments are much too detailed to include herein, but several of his statements can be used to illustrate the lack of notification by the EPA.

Lynch notes:

> It is fortunate that the Department of the Interior apparently felt strongly enough about the subject rule (new regulations) that they followed their standard procedures and informed mining claimants of record about the EPA publication (of the Final Rule) in the Federal Register, and the comment period. This notice, however, was not received by me until July 12, 1988 (after the effective date of the regulations), when I returned from a week and a half at my mining operation. I reviewed the subject material and discovered the comment period was not specified.

On July 19, Lynch contacted the National Technical Information Service to obtain "four major documents" listed by the EPA and was told by the NTIS that they had no such documents available. He contacted the EPA the same day and was informed that the Economic Impact Analysis document was being bound at that time and had not yet been sent out for distribution. The EPA could not tell him the status of the Development document, which contained the economic data that estimated the costs of constructing and operating the control equipment required by the new ruling (this is 12 days *after* the final rule went into effect). Lynch received copies of two of the requested documents on Aug. 2 with a note saying, "The Development Document has not [yet] been printed for distribution.... It should be available soon. When it is available, we will send you a copy."

As a result, Lynch had to submit his analysis and comments on the EPA regulations without having an opportunity to review the Development document. In his final statement, he requested that he be allowed to submit additional comments on the final rule after he had an opportunity to review the missing information.

Expressing his understandable frustration, Lynch's closing comment in his letter to *CMJ* stated the following:

> It is now all set up so that the Sierra Club or anyone else who thinks you are muddying the waters of the United States can close you down, and put you in jeopardy of a significant penalty. I hope someone, somewhere, can do something about this one.

In a telephone call to *CMJ*, one caller expressed concerns over the following:

- There was inadequate (or nonexistent) notice of the proposed (finalized?) regulations.
- The new rules may not apply to operations processing under 20 cubic yards of gravel per day or 15,000 cu. yds. per year (some confusion on this item).
- No new placer mines are to discharge waters into streams.
- The Final Rule was published over the objections of the U.S. Bureau of Mines and the Department of the Interior.

One *very* significant point to be made here is the lack of availability of the supporting documents referred to by the EPA in the information published in the Federal Register in May 1988. Under the Section II heading "Scope of this Rulemaking," the EPA states:

> This final regulation, which was proposed on November 20, 1985 (50 FR 47982), establishes effluent limitations guidelines and standards for existing and new source gold placer mine facilities....
>
> This final regulation is supported by four major documents, three of which are available from the National Technical Information Service. Analytical methods are discussed in 'Sampling and Analysis Procedures for Screening of Industrial Effluents for Priority Pollutants.
>
> EPA's technical conclusions are detailed in the 'Development Document for Effluent Limitations Guidelines and Standards for the Gold Placer Mine Subcategory of the Ore Mining and Dressing Point Source Category.
>
> The Agency's economic analysis is found in 'Economic Impact Analysis of Effluent Limitations Guidelines and Standards for the Gold Placer Mine Subcategory.

We might be inclined to wonder just exactly why most of these supporting documents were still unavailable to the general public (and to placer miners) at a point in time *well after* the effective date of the final regulations being implemented by the EPA. How can one possibly be expected to make an informed comment on regulations for which complete information is not available, or to comment effectively on new regulations that have already gone into effect—particularly where advance notification was deliberately withheld?

6
Grassroots U.S. Mineral Exploration Declines Sharply

THE SEARCH FOR NEW MINERAL deposits at the grassroots level in the United States has been undergoing a gradual decline over the past year or so and, in the past few months, exploration activity has dropped sharply. At present, most of the large and small mining companies are concentrating almost exclusively on identified exploration targets and those mining properties that already have defined ore deposits. Only a few mining companies are still examining new prospects, and they are only interested in acquiring outstanding properties for exploration projects.

The primary reasons behind this major drop in new minerals exploration activity appears to be a direct result of these factors: (1) mineral commodity prices being at low levels, which lowers profits and reduces capital available for exploration and acquisition; (2) stringent regulations, which produce long time delays, major increases in expenditures and a reduced potential for making a profit; (3) pending legislation and additional new stringent regulations, which make business planning impossible, increase the risk associated with investing capital in mining ventures, and pose a constant threat of imposing unlimited liabilities on all mining activity, and (4) a hostile business environment, in which anti-mining activists, the news media and certain politicians are mounting constant attacks on the minerals-producing sector of the economy.

The abovementioned problem areas have also been considerably magnified by the penalizing nature of regulatory enforcement. Most regulatory agencies are now emphasizing the use of civil and administrative fines and penalties as a means of forcing everyone to comply with their rules and regulations, rather than instituting programs utilizing a friendly and cooperative approach to help business and industry correct any problems. The idea of "sending a message to potential violators through the zealous prosecution of all 'offenders,'" imposing the maximum penalties allowed by law without regard to the seriousness of the violations (or whether the violations were intentional, unintentional or accidental), has grown to the point where it now dominates the regulatory mentality.

The increasing severity of these problems over the past two decades has been sending signals to miners for some time now, and the message has gotten through: mining is no longer welcome in the United States. As a direct result, most major mining companies are shifting their exploration and development efforts to foreign countries while cutting back on their operations in the U.S. This is, after all, a practical business decision: why try to operate in an unfavorable business climate when many foreign countries are encouraging mineral development and welcoming U.S. companies with open arms?

Small-scale miners and prospectors, who are directly responsible for the vast majority of new mineral discoveries, are being impacted even more severely by this situation. Where is the incentive to prospect for mineral deposits when it is increasingly likely that you will never be able to mine them, or realize anything worthwhile for your efforts? Especially if small-scale miners find themselves spending most of their time trying to cope with the overwhelming number of overly stringent new laws and regulations, as well as constantly fighting with regulatory agencies. As a result, fewer and fewer small-scale miners and prospectors are willing to continue the fight against almost impossible odds, let alone keep up their enthusiasm for prospecting for new mineral discoveries.

Mineral exploration is the life-blood of the mining industry, and it is being choked to death by legislative overkill, a regulatory nightmare, environmental extremism and radical anti-mining groups. Without the discovery of new mineral deposits, there will be no new mines. As known ore reserves are mined out, there will be nothing to replace them. In short, if you kill mineral exploration in the U.S., then the U.S. mining industry is living on borrowed time.

An examination of the available information reveals that new, grassroots mineral exploration by mining companies in the United States has declined by roughly 80% over levels seen just two years ago. In addition, U.S. mining companies have laid off at least 200 exploration geologists within the past six to eight months. Several major companies have closed down their exploration offices in the U.S., and a number of smaller exploration firms have gone out of business. Most of the independent consulting geologists and mining engineers are unable to find any work—no matter how hard they look—and quite a few of them have been forced into seeking employment in other types of businesses or going to work for mining firms located in foreign countries. Almost all of the remaining exploration firms and exploration departments of major mining firms have made drastic cutbacks in personnel, have cancelled new mineral exploration programs, have stopped looking at mining property submittals for grassroots prospects, and are concentrating solely on acquiring properties with known ore reserves or conducting exploration on existing projects with proven targets. In addition, investment capital for mineral exploration has evaporated.

At the same time, however, mineral exploration activity in foreign countries is booming. Many of the major mining companies that have stopped all exploration programs in the U.S. have actually increased their mineral exploration efforts overseas. Several of the major mining firms have been selling off their U.S. assets at the same time that they have been actively acquiring new mining properties in foreign countries. So, it is fairly obvious that this is not just a temporary down cycle in mining activity in this country; it is becoming quite evident that there is an increasing movement of U.S. mining companies to overseas locations.

Unless there is a major positive shift in U.S. minerals policy in the very near future, this trend towards moving offshore will be permanent. In just a few years, U.S. mineral production will decline sharply also and eventually stabilize at insignificant levels. It naturally follows that imports of mineral commodities from foreign sources will rise dramatically as a result, and the U.S. will become almost totally dependent upon other countries for its minerals supplies.

Quite a few people have been predicting dire consequences for the U.S. economy if more onerous federal and state legislation is enacted, if the trend towards regulatory overkill continues, or if the regulatory agencies' policy of "fine 'em and jail 'em" is allowed to persist. Well, the process has already begun to take effect, and the Sierra Club objective of "Mine Free by '93" looks more likely with each passing day.

Unless miners stand up and fight this insanity with every means available, the final outcome is a foregone conclusion.

7
Miners See Light at the End of the Tunnel

WHILE THE CONGRESS AND MANY state legislators have not yet packed up and gone home as of this date (early October), it is already evident that miners will not be put out of business this year. In fact, it now appears that the future prospects for mining in the U.S. may be brightening somewhat. Although there is still a long way to go before the anti-mining factions have been brought to a complete halt, a little light is beginning to appear at the end of the tunnel.

Considering the large number of critical issues involved, mining will come out in reasonably good shape this year.

However, miners should not derive a false sense of security from a successful defensive campaign against environmental extremists and anti-mining groups, because they are continuing to mount massive campaigns to eliminate all types of natural resource development in the United States. In other words, a win when you fend off an attack by an opponent is not anywhere near the win achieved when you are on the offensive and subject your opponent to a resounding defeat. In this case, our enemies' forces are still intact and on the offensive—so mining will continue to be subjected to attacks by certain politicians, by environmental radicals and by anti-development groups.

There are some definite indications, though, that the power exerted by radical extremists in the environmental movement reached its peak sometime around early summer of this year, and that the nation's environmental mania is beginning to show signs of an irreversible decline. The primary reason for this phenomenon lies in the obvious excesses and extremes to which the radical elements have resorted to advance their own agenda. The totally irrational and almost unbelievable demands being made by supposed "protectors" of the environment are now becoming evident to members of the general public, and they are now beginning to say that extremism in environmental matters is no longer acceptable.

There is some evidence that the radical environmentalists have also recognized the beginning of a trend towards more moderate and conservative approaches to environmental matters. Most of these groups have recently begun to reduce their direct attacks on specific natural resource industries and have shifted their efforts toward utilizing other available means to accomplish their desired objectives. As an example, the wide-open opportunities presented by some ill-conceived provisions in the Endangered Species Act are now being used effectively to shut down all types of natural resource development and utilization on massive tracts of both public and private lands. The Federal EPA's incredible "wetlands" definition and totally unrealistic standards for a multitude of particulate and chemical emissions have presented anti-development interests with unlimited opportunities for rampant litigation and obtaining massive funding for their activities by the semi-legal blackmailing of development firms and industrial corporations.

So, while miners have made some positive gains in the legislative arena and are beginning to see some signs that the power wielded by radical preservationists is waning, the war for survival is not yet over. This is especially true in the regulatory process, where far-reaching decisions are being made with little thought being given to their eventual impacts on society and the nation's economy. This is not just a problem for miners; producing enterprises in the U.S. are being overwhelmed by an ever-expanding bureaucracy, stringent and inflexible laws, short-sighted policies, ever-increasing taxes and fees, rampant litigation, and totally illogical and ill-conceived over-regulation. In fact, most of the nation's producers and builders are being strangled by the plethora of new rules, laws and regulations while our leaders are frantically encouraging the consumption of products by the so-called "consumer" in order to keep the economy from falling into a severe recession or depression.

Considering all of the somewhat depressing factors mentioned above, can we find something positive in this undeniable mess we're in? Just where is this light at the end of the tunnel? For one thing, miners are in a much better position now than they were at the beginning of this year, when it appeared very likely that the U.S. Congress would put us out of business in one fell swoop. Well, you might ask, why isn't it just as likely that they will put us out of business next year? There are several good reasons why mining should receive better treatment in the U.S. Congress next year (for the next few years, in fact), including: (1) we have gained a considerable amount of backing in the U.S. Senate, which resulted from some very intensive efforts put forth by the mining community and our supporters; (2) 1992 is an election year, and most politicians are going to be very careful when choosing which controversial issues they are going to back (if any); (3) legislators are going to be looking for ways to help increase economic activity, rather than decrease it; (4) the extremist elements in the environmental movement are beginning to lose some of their (undeserved) credibility with lawmakers, and (5) we will still be dealing with the same U.S. senators we dealt with this year, and we are continuing to gain more backing in the U.S. Senate.

It can now be stated, with a good degree of certainty, that almost all of the major mining issues in the U.S. Congress have been defeated, to wit: Sen. Dale Bumpers' S. 433 to abolish the 1872 Mining Law is gathering dust in the drawer of a subcommittee of the Senate Energy and Natural Resources Committee, from which it will not see the light of day for the remainder of this year; efforts to rewrite Rep. Nick Rahall's H.R. 918 appear to have stalled and, even if the bill is ramrodded through the House Interior Committee and passed by the full House of Representatives, it would most certainly not make it out of the U.S. Senate; Rep. Peter DeFazio's H.R. 2614 will never surface in Rep. Rahall's subcommittee and would never get anywhere if it did; and H.R. 1096, the so-called BLM reauthorization bill that would make the Bureau of Land Management into an environmental agency, is effectively dead in the U.S. Senate.

In addition, the $100 mining claim holding fee has been deleted from the FY1992 Interior Appropriations bill (H.R. 2686) and is considered extremely unlikely to resurface, and the mining patent moratorium proposal in the same bill was defeated 47-46 on the floor of the U.S. Senate. According to congressional staffers, none of the abovementioned legislative measures will resurface this late in the 1991 session of Congress, especially when considering the plethora of major issues that remain to be resolved before Congress adjourns.

It should be pointed out that this rather nebulous win for miners was achieved only through the tremendous efforts exerted by both small-scale miners and the mining industry and the critical backing provided by a number of our friends in the political arena—particularly a few outstanding U.S. senators from the western states.

Another very positive development in the fact that the federal administration and an increasing number of the members of Congress have realized the problems inherent in overregulation, and have recognized that American business is being handicapped (actually hamstrung) by the multitude of laws, rules and regulations—many of which are now being publicly recognized as being ill-conceived and overly stringent. In other words, there is a chance that the trend toward a regimented and government-controlled society might slowly be replaced by a more rational approach to government. An important example of this potential lies in the increasing recognition of private property rights and just compensation on "takings" issues. There appears to be a better chance to bring about change in this area now than was possible in the past two decades. However, it will take a lot of hard work and it won't be easy because of the resistance to change exerted by an entrenched bureaucracy, but it must be accomplished because overregulation alone could put miners (and everyone else) out of business.

In summary, 1991 had the potential for being the worst year ever for mining and other natural resource industries—but it wasn't. The massive legislative and regulatory campaigns mounted by anti-mining politicians, environmental extremists and a biased media, along with the support provided by a misled and misinformed general public, peaked out this year—and they did not achieve their major objectives. They will undoubtedly continue with redoubled efforts, using every available means to advance their causes, but it is highly unlikely that they will ever again achieve the level of power and influence that they possessed earlier this year.

Even though a glimmer of light has now appeared at the end of the tunnel, we cannot slacken our efforts. As has been quite evident throughout all of this year, there is still a critical need for continued unity of purpose, cooperation, organization, communication and action from the individual and collective members of the nation's mining community. Now is the time to go on the offensive instead of continually falling back in a defensive posture. Miners and other natural resource users can no longer afford the luxury of sitting back and letting "the other guy" try to take care of these problems, because experience has shown that it just won't work. So, get involved!

8
U.S. Mineral Exploration Is Being Decimated

JUST A FEW YEARS AGO, everyone was talking about the new mining "boom" in the United States and the positive outlook for the national economy. In a relatively short span of time both have undergone a rather sharp decline, and current conditions do not bode well for the future—particularly in reference to mining. It is beginning to appear that the new mining boom will be referred to historically as "the last mining boom."

Although all of the data is not yet available, there is sufficient information on hand to show that U.S. mineral exploration, absent any significant change in federal policy, is on a permanent decline. A rough estimate last year indicated that grassroots mineral exploration had dropped about 80% over a 3-year period, and this year the indications are even worse. Exploration staffs and budgets have been cut to the bone or eliminated entirely for many mining companies, more exploration firms have shut down or gone out of business, small-scale mining operations are disappearing, more ore deposits have been mined out than there have been new deposits discovered to replace them, very few independent miners and consultants are left in the business, and now we are seeing a sharp reduction in mining claim activity as well as a sharp drop in the number of claims being held by both individuals and companies.

In Nevada alone, which has the largest number of mining claims of any state in the U.S., the number of new claims filed had dropped 69% over a period of 4 years (to September 30, 1991, the end of the federal fiscal year). The number of new claims filed in first quarter 1992 had declined 44% from the number filed in the first quarter of 1991, so the trend is continuing. The total number of Nevada mining claims (new claims plus existing claims) to the end of the state's fiscal year (July 30) appears to have dropped 17% from 1990 through 1992.

On a national basis, the BLM recently estimated a total of 1.2 million active mining claims in the U.S. The Office of Management and Budget has estimated 976,000 claims nationwide for its budget estimate prepared for the projected revenues the OMB expects from the $100 mining claim holding fee. The most recent BLM figures for the total number of active claims in the first half of FY1992 (from October 1, 1991 to April 1, 1992) shows a national combined total of under 700,000 claims. Since well over 90% of the Affidavits of Labor and Notices of Intent to Hold are filed between October and December 30 of each year, it would appear this number is fairly close to the actual figure.

These figures would indicate an approximate 42% drop in the total number of active claims nationwide (from 1.2 million claims down to 700,000). This would also indicate a drop of 28% from the OMB's estimate (976,000 down to 700,000), and the OMB had estimated their figure *after* subtracting the estimated claims that would be dropped as a result of the imposition of the $100 holding fee.

The total number of new mining claims filed with the BLM in a national basis each year has also been falling sharply: from 1989 to 1990, new filings had dropped 38% (from 158,511 down to 97,455); from 1990 to 1991, new filings dropped another 20% (from 97,455 down to 78,362); and, from 1989 to 1991, new filings had dropped over 50% (from 158,511 down to 78,362). Between 1991 and 1992, data shows an additional drop of about 16% in the number of new claims filed in the first two quarters of 1992 (federal fiscal year) as compared to the first two quarters of 1991 (from 43,756 down to 36,876). Over the past 5 years, from 1988 to April 1, 1992, there has been an overall 60% drop in the number of new claim filings nationwide.

Now, what does all this mean? It means that while the new mining boom was still reaching its peak, mineral exploration was already in a severe decline. It is now apparent that U.S. mineral exploration has been reduced by about 90% since it peaked in late 1988 to early 1989.

And now, let's examine the factors that are directly responsible for this decline: (1) stringent overregulation, which produces long delays in permitting, creates major increases in expenditures, and reduces the potential for making a profit; (2) pending legislation and new stringent regulations at the federal and state level, which make business planning impossible, increase risks for investing capital in mining projects, pose a constant threat of imposing unlimited liability on all mining activity, and present a very real potential for shutting down mining at any time; (3) an extremely hostile business environment, in which anti-mining activists, the national news media and certain members of Congress constantly attack the minerals-producing sector of the economy, and (4) low mineral commodity prices, with the consequent reduction in profits and less capital available for mineral exploration and acquisition. No one in his right mind will continue to spend time and money on mineral exploration when it is increasingly likely your own government will never allow you to mine your discoveries.

If no new mineral deposits are discovered, there will be no new mines. As known ore reserves are mined out, there will be nothing to replace them. Therefore, if you kill exploration in this country, then the U.S. mining industry is living on borrowed time.

9
Congress Is Killing America's Economy

THOSE OF US WHO WORK in natural resource industries (mining, ranching, timber, fisheries, etc.) have known for some time that the majority of the members of the U.S. Congress want to shut us down, permanently. After years of being subjected to anti-development propaganda in the national news media and intense pressure from radical environmental extremists, most of our elected leaders now firmly believe that the nation and nature would be much better off if we were eliminated.

Up until recently, most of those employed in other types of businesses believed they were somehow exempt from the extreme overregulation and micromanagement imposed on the natural resources sector by government. Now, however, almost all businesses in the United States are finding themselves stretched to the limit and unable to engage in any productive enterprise without the interference of big government. They are literally being micromanaged to death.

Despite the recent open acknowledgment by the federal government that unnecessary overregulation was stifling the American economy, the bureaucracy continues to pump out an overwhelming number of new and costly laws, rules and regulations. The current preoccupation with environmental issues has accelerated this process to the point where it is now seriously impacting every U.S. citizen—except the members of Congress, that is, because they continue to exempt themselves from the laws and regulations which they impose on the rest of the population.

Although government has consistently shown that it is totally incapable of managing itself, it persists in the belief that it can better manage all businesses and each individual's personal life. And, even in an election year when we should have had some breathing space, this trend towards total government control of everything and everybody continues at an accelerated pace.

Unemployment in the United States was recently reported at an eight-year high (7.5%), and the government is only counting those people who are actively seeking employment. There are millions more who have, at least temporarily, given up hope of finding work and are, therefore, no longer included in the statistics. At the same time, there are literally hundreds of new bills being considered in the U.S. Congress that will throw thousands more out of work (maybe even millions) and produce astronomical price increases for just about every product produced and sold in this country. There is absolutely no economic balance in the legislative process at this time. And, for almost all regulatory purposes, economic considerations are nonexistent.

So, what solution to the current economic difficulties is being offered by the government bureaucracy? They are actively encouraging consumers and businesses to accumulate more debt, thereby supposedly producing an economic rebound. Strange way to do business, isn't it?

While the U.S. Congress continues to ensure the constant accumulation of record-breaking annual federal deficits (which is mortgaging each and every American citizen's economic future), they actively encourage the private sector to accumulate more debt in order to generate more revenues to the U.S. Treasury which, in turn, allows the government to spend even more of our money. In the interim, government continues to impose ever-increasing "fees" on every conceivable activity for the privilege of doing business in the United States.

What is this doing to the nation's productive businesses? Many of them are either going bankrupt, closing down, or moving out of the country. We are literally exporting our most productive enterprises and millions of jobs while, at the same time, ensuring that millions of Americans will be unable to find a job. You can't blame the businesses for leaving the country: If your own government creates conditions under which your business can no longer function in this country, you either close your doors or go to some other country where they welcome your business with open arms.

The government bureaucracy has grown to such a massive size that the American people are now incapable of supporting it. As a result, it now relies on a constant influx of foreign investment capital just to keep operating from day to day. If just one of the U.S. Treasury's quarterly debt offerings did not attract any foreign investors at all, the entire government financial structure would collapse. The only remaining choice the Treasury would have would be to print more currency, which would have a disastrous effect on the U.S. dollar's exchange rate on the international markets and produce massive price inflation in the United States.

Over two years ago, several respected economists reported that if foreign investment capital in the U.S. economy disappeared overnight (and this is not that unlikely), then the Gross National Product would immediately decline by 25%. The situation is much worse now. In other words, our government has gotten us into such bad financial shape that we are almost totally reliant on foreign money for economic survival. And, the U.S. Congress is doing absolutely nothing about this dangerous situation. For that matter, the Congress is rapidly making the situation much worse. When our leaders are racking up our national debt at the rate of half a trillion dollars or better in one year, nothing can be done to even begin to correct the problem until our government's profligate spending is brought under control.

So, our current economic difficulties have been created entirely by the government bureaucracy and the U.S. Congress, and they now expect us to bail them out by increasing our debt load. At the same time, Congress is still doing everything it can to kill America's economy—which means we cannot borrow enough money to bail our government out of this mess anyway.

10
Clinton Launches Attack on American West

WHEN PRESIDENT CLINTON REVEALED the administration's plan for the nation's economy before Congress on February 17, 1993, he released a document entitled *A Vision of Change for America,* in which the federal budget was generally outlined. Despite the recent propaganda expounded by government officials and the media, an examination of Clinton's plan shows that Westerners are being asked to support a number of "revenue enhancements" that will practically decimate many of the West's most productive basic industries and wipe out the economic base of many rural communities. In effect, our citizens are being asked to accept higher unemployment, severely reduced economic activity, and the loss of significant state, county and local revenues. For the purported purpose of strengthening the overall U.S. economy and meeting the administration's environmental goals, the public land states in the West are to be sacrificed on the altar of utopian idealism.

For example, on page 78 of the *Vision* document it is stated:

Interior/Permanently extend hardrock mining holding fees. There are over one million hardrock mining claims on Federal lands operating under the 1872 Mining Law. Hardrock minerals include gold, silver, lead, copper, zinc and numerous other minerals. The 1872 Law requires claimants to annually perform $100 worth of work to develop and maintain their claims. Claimants must pay only a filing fee to the Federal government for the right to mine this land. This proposal would permanently authorize charging a $100 per claim holding fee on all hardrock claims on Federal lands (extending a fee enacted for 1993). The claimant would be relieved of annual work requirements, which should increase his or her flexibility on timing the development of claims. It will also reduce unnecessary ground disturbances to satisfy current law. The proposal would increase revenues to the Treasury by an estimated $80 million per year, after covering the costs of administering the entire hardrock mining program, including environmental compliance. It would generate estimated revenues of $320 million in 1994-97.

Let's examine this proposal carefully. There are not over 1 million claims—the actual figure is much closer to 726,715 claims *before* miners dropped any claims in 1992 and *before* the imposition of the $200 per claim on August 31, 1993. Recent mining industry estimates show about 50% of the existing claims being dropped because of the holding fee requirement, which leaves around 360,357 actual valid claims by late August of this year. This would produce a revenue estimate of around $36,335,700 from which BLM funding of $17,430,000 is allocated. Total revenues to the U.S. Treasury would equal about $18,905,700 for an annual revenue shortfall of roughly $61,294,300. Because the 1994 fees are paid in advance, the same figures would apply to FY1993-94, although the government would receive both years' revenues at the same time. So, over the period of FY1994 through FY1997, the total revenue shortfall would equal $245,177,200. This is, of course, assuming no further drop in mining claims over the period.

Who does the math for these guys, anyway? The projected $320 million would be spent by Congress because it is allocated to programs in the new budget. Therefore, the mining claim holding fee would actually contribute over $245 million to the federal deficit while eliminating much of the remaining U.S. mineral exploration. Insofar as "flexibility on timing the development of claims" goes, who can afford to develop minerals if almost all your potential investment capital is being paid in cash to government?

And then we have the double whammy. Still on page 78 of the same document is stated the following:

> Interior/Institute hardrock mining royalties. Establish a "12.5 percent royalty on the gross value" of the hardrock minerals extracted from mining claims on public lands. There are over one million hardrock mining claims on Federal public lands operating under the 1872 Mining Law. Hardrock minerals include gold, silver, lead, copper, zinc and numerous other minerals. The 1872 Law was one of many laws intended to encourage the settlement and development of the West. It allows miners to prospect, make claims, and extract minerals from Federal lands for the cost of filing a claim. There is no current authorization to charge a royalty on hardrock minerals privately extracted from public lands. Laws enacted early in this century provide for Federal leasing and collection of royalties from oil, gas, coal and certain other minerals extracted from Federal lands. Hardrock mining, however, remains under the rules of the 1872 Law. The new royalty would be phased in over three years. The time necessary to set up and administer the royalty in the most effective way would delay initiation of royalty collection until 1995. Receipts from hardrock mining royalties would be

shared with the States where the mining occurs. This proposal is expected to pay for the costs of enforcement and collection. It would generate estimated revenues of $471 million in 1994-1997, including $277 million in 1997.

Our federal government is, apparently, living in Fantasy Land. These people actually think that you can impose huge increases in fees, taxes, royalties and other charges on American businesses, calculate the resultant revenues to the Treasury *based on the current gross value of production,* and spend those projected revenues on government programs. For some strange reason they do not seem to think that the imposition of higher costs on business would affect the value and rate of production (let alone the employment and economic activity generated by business).

Now, if you impose a 12.5% gross production royalty on mining, where commodity prices are set in international markets, you effectively reduce the value of each particular commodity by that percentage. For instance, gold valued at $330 per troy ounce would be reduced to an effective value of $288.75; silver at $3.70 per ounce is reduced to $3.24; and copper at $.98 per pound devalues to about $.86 per pound. *Mines are already closing down because of low commodity prices, so what happens when you effectively reduce those commodity prices another 12.5%?*

What about low-grade ore reserves? Same thing: if you lower profitability of a mining operation then the economic cutoff grade rises, thereby forcing the mine to leave more mineral-bearing ore reserves in the ground and shortening mine life considerably. For primarily low-grade mines, you force them to shut down immediately. This would, in turn, produce much higher unemployment, reduce economic activity and, overall, produce a rather sharp drop in federal, state and local tax revenues.

Now, let's look at a few other aspects of Clinton's plan to strengthen the U.S. economy: the federal government plans to hit ranchers in the West with large increases in grazing fees; they plan to drastically cut back on timber sales and increase fees on the remaining sales; they will impose a water surcharge on farmers and ranchers who use water from federal water projects, and they plan significant increases in fees (and imposition of new fees) for recreational use of the public lands.

All of the fees and royalties listed above are shown in the budgeted plan as "elimination of subsidies" and "cuts in federal spending." Since *all* of the costs associated with mineral exploration, development and production are (and always have been) borne entirely by the mining industry, it would be interesting to find out exactly how the government has been "subsidizing" America's minerals industries. For that matter, government has been actively subsidizing foreign minerals development, at the expense of U.S. miners and taxpayers, for many years.

Overall, the president's budget proposals constitute a virtual wish-list of the nation's environmental movement, and these groups are hailing the plan as a sign of fundamental change in federal land-use policies. The way it looks now, federal land management policy has changed from one of "multiple use" of the public lands to one of "no use," and the West is slated to become a vast, natural park for visiting Easterners.

PART II

Mining and the U.S. Forest Service

"Is it any wonder that most rabid environmentalists are seeking to place more of our public lands under the control of the U.S. Forest Service?... The manner in which the regulations have been constructed and the specific wording used make them appear to have been carefully designed by environmental activists, such as the Sierra Club."

11
USFS Mineral Examinations Violate Claimant's Rights

Editor's Note: See related articles, 21 ("USFS Responds to Author's Article on Claimant's Rights Issue") and 22 ("Author's Response to USFS Letters on Claimant's Rights Issue").

DID YOU KNOW THAT the U.S. Forest Service allows mineral examinations on mining claims in the national forests without notification of the mining claimant that an attempt is being made to invalidate their claims? Were you aware that Forest Service mineral examiners may ignore the mineral examination procedures required by the Department of the Interior—even though these regulations govern mining claim contest actions that fall under the jurisdiction of the Department? If not, then you probably did not know that the Forest Service says it doesn't even have to make an attempt to notify a mining claim owner before the agency initiates action to invalidate your claims.

Consider this scenario: A mining claimant in Nevada received notice in May 1986 that his claims had been found to be invalid by a U.S. Forest Service mineral examiner, and that a mining claim contest action had been initiated by the Bureau of Land Management at the request of the Forest Service. This was the first contact with the mining claimant, so he had no idea that a validity determination was even being considered. He then filed an appeal to the contest action through an attorney. In late February 1987 a copy of the mineral examination report was finally received by the claimant. Imagine his surprise when he learned that the mineral examination of his mining property was undertaken in late 1983 and completed in December 1985—without any attempt at notification having been made by the Forest Service.

An administrative law judge held a hearing on the matter in October 1987, during which the mineral examiner testified that no attempt was made to notify the claimant because it was not required. In addition, when asked why the mineral examination procedures were not followed, he stated that the Forest Service is not bound by those regulations. The mineral examiner went on to say that he asked an equipment operator at the site (who worked for a trucking company and had no mining expertise) to point out the areas where mineral "discovery" had been made for purposes of ascertaining the examiner's sampling locations, and noted that this individual was used as the claimant's "authorized representative" for purposes of the examination process. He also acknowledged that previously compiled mineralogical and geological data on the properties was not examined or even requested, and that the claimant had not been provided an opportunity to participate in the examination process.

Despite these obvious violations of due process and the accepted mineral examination procedures, his testimony and report were utilized by the government as the basis of their case and this information was duly entered into the record. The case is still pending, and a decision will not be rendered for another six months. In the meantime, the claimant has paid thousands of dollars in expenses, has not been allowed to develop his mining property for one and a half years, and cannot proceed with further mining activity until a final decision is handed down. The Forest Service has not allowed any operations on the site since the contest action was initiated.

By the way, the claim owner *has* made a valid discovery of valuable minerals on the property and can prove it—but the Forest Service refuses to consider the evidence because their mineral examiner did not "discover" the minerals by himself. In addition, the property is located within one mile of a major known gold deposit and within six miles of one of the largest gold-producing mines in the United States—in a district where *all* of the available public lands are covered by many *hundreds* of mining claims.

This is an actual case, and as you might have noticed, I have avoided using the names of the individuals and properties involved because the case is still pending. I can, however, describe the information I obtained while researching the regulations pertaining to USFS mineral examinations.

First of all, there *aren't* any regulations governing Forest Service mineral examinations. As one Forest Service spokesman put it: "The various handbooks, procedures and regulations are merely guidelines, and we are not required to follow them." Another Forest Service regional supervisor said, "We normally notify the claim owner of pending mineral examinations, but we are not required to do this." A BLM spokesman said, "We ask the Forest Service to follow our mineral examination procedures, but they often do not. They say that they're only required to follow their own regulations." Which, in this case, apparently do not even exist.

The U.S. Forest Service obtains its authority to conduct minerals examinations on national forest lands through a "memorandum of agreement" with the Bureau of Land Management, which retains the final authority for minerals management on the nation's public lands. This "memorandum" is a rather ambiguous document that provides the USFS with authority to perform mineral examinations but *does not* require that agency to follow the appropriate procedures outlined for the conduct of same by the Department of the Interior. This gives the USFS "discretionary" powers that enable its mineral examiners to totally disregard the claimant's right to notification and participation in the mineral examination process, and allows them to inspect and evaluate the property in any way that they see fit.

This, in turn, provides the Forest Service with an opportunity to stack the deck against any mining claim owner whom they might wish to evict from lands under their jurisdiction. Mineral examiners are allowed to pick and choose which "guidelines" they observe, or even to ignore them entirely, *while the government bases its mining claim contests almost entirely upon the mineral examiner's report!* This place the U.S. Forest Service in a unique position: They are allowed to operate outside of the regulatory framework that provides the mining claimant's "due process" rights.

Imagine this, if you will: You own mining claims in an area that is of special interest to the Forest Service or, perhaps, you have a conflict with certain USFS personnel. A year or two in the future you receive a notice that your claims have been examined and have been found to be invalid by the Forest Service (notification *after* the fact). You then have 30 days in which to retain an attorney and file a reply (appeal) to the claim contest, and have to wait about two years before a hearing is scheduled on the matter. In the meantime, the Forest Service cancels your plan of operations and/or refuses to allow you to work your mining property during this time period. When the hearing is finally held (and even if you win), you still have to wait another six months or so before a final decision is reached—during which time you are also not allowed to mine your claims. And, remember, you weren't even informed that a mineral examination was in progress and you weren't given an opportunity to participate (or a chance to point out your mineral discoveries). How would you feel?

This is not a hypothetical case. Unless and until some form of control is placed on USFS mineral examinations, they will have the authority to ignore the relevant regulations and proceed to initiate a validity determination at any time—*without notifying the claimant of their intentions.* And, having obtained a significant advantage over the mine owner by eliminating him or her from the process, the Forest Service will seldom lose any appeal of their original determination.

Incidentally, I have obtained copies of the BLM and USFS procedures outlining the conduct of mineral examinations and studied them. I also went over the documents with an attorney, talked to three regional Forest Service supervisors, discussed the issue with two USFS mineral examiners and two BLM mineral examiners, and questioned two individuals in the USFS legal department. They *all* stated that the USFS is *not required* to notify mining claimants of pending validity examinations but added that this procedure is "normally" followed. They also agreed that there were no regulations governing USFS mineral examinations—only "guidelines" that are optional at the *discretion* of USFS personnel. Is it any wonder that most rabid environmentalists are seeking to place more of our public lands under the control of the U.S. Forest Service?

Let's go a bit further on this. As many miners now know, the definition of mineral discovery on mining claims has been "refined" by the "prudent man" and "marketability" tests. Under these standards, an ordinary person (not a miner) must be justified in the further expenditure of time and money with a reasonable prospect of developing a valuable mine by making a discovery of valuable minerals in such quality and quantity that a "prudent" person would make a similar judgment. In addition, the minerals must meet the "marketability" standard, which generally means that it must be proven that the minerals can be mined, processed, transported and sold at a profit. Very few mining claims would meet these stringent requirements, particularly those in the early stages of exploration and development.

Now, visualize a person with sufficient power and authority entering upon the scene—particularly one who *wants* to invalidate certain mining claims. Then give this person the additional authority to ignore the regulations governing mineral examination procedures, and allow them to inspect and evaluate a mining property without notifying *anyone* that a validity determination is in progress. Just what type of report would you expect from this person, and what chances would you give the mine owner for successfully appealing the determination in an administrative hearing? I think you might have guessed the probable outcome.

You see, the burden of proof rests with the mining claimant whenever the validity of a mining claim is contested by the government. Evidence presented by the mineral examiner and government witnesses is presumed to be factual, and it must therefore be countered by a preponderance of verifiable evidence from the persons appealing the determination. Only "expert" witnesses are allowed to testify for the defense, and they may only give testimony on matters of personal knowledge that lie within their particular area of expertise. In an analogy with case law, the miner is judged guilty until proven innocent. As can be seen, a defective mineral examination can provide the government with an overwhelming advantage—particularly if the mining claimant is denied the opportunity to participate in the examination of his property. In addition, how can the claimant gather evidence to support his claim if the Forest Service will not allow him to develop the mineral discoveries on the property?

This type of "covert" operation by the Forest Service must be brought to a halt immediately, and some type of controls must be instituted to correct deficiencies in the USFS mineral examination procedures. Without some form of definitive regulation, a few unscrupulous individuals can make a mockery of the mining laws. The existing Memorandum of Agreement between the BLM and the USFS should be scrapped until a more acceptable arrangement can be devised. If the U.S. Forest Service is to be allowed to continue with mineral examinations, then they should be bound by the same regulations governing mineral examinations performed by the Bureau of Land Management. Nothing less is acceptable.

I strongly suggest that all miners contact their congressman and state legislators, and express their concerns over this matter. It is the only way that this situation might be corrected.

12
USFS Responds to Author's Article on Claimant's Rights Issue

Editor's Note: The following "letters to the editor" were sent to the *California Mining Journal* in late 1987 or early 1988 in response to author Dave W. Parkhurst's article titled "USFS Mineral Examinations Violate Claimant's Rights" (article 20 in this compilation) printed in the December 1987 *CMJ*. The letters were published in the January 1988 issue of the magazine. The writers of the first two letters, E.R. Browning and Denton W. Carlson, were with the U.S. Forest Service, Minerals Area Management, at the time the letters appeared in the *CMJ*. The third letter was written by the president of the Mother Lode Miners. The article that prompted the letters of response, combined with the letters themselves and the subsequent response by Dave Parkhurst (article 22 in this volume) are illustrative of the relationship and interactions between miners and the federal regulatory agencies, specifically the USFS, during the time the article and letters were originally published in the *CMJ*—as viewed from the perspectives of the individual writers speaking for the mining community and from those of the USFS representatives.

Letter from E.R. Browning:

An article in the December 1987 issue of the California Mining Journal entitled "USFS Mineral Examinations Violate Claimant's Rights," raises some questions concerning U.S. Forest Service (USFS) procedures for conducting mineral examinations of mining claims on reserved public domain lands. We would like to respond to this article and set the record straight on several points.

A 1957 Memorandum of Agreement between the Forest Service and the Bureau of Land Management (BLM) establishes the work procedures governing action on claims for lands within national forests. Under this memorandum, the Forest Service makes mineral examinations to establish the validity of unpatented mining claims or mill sites in response to mineral patent application, to verify the presence of valid existing rights to mining claims in withdrawals, and to determine validity in other situations when such action is deemed to be in the public's interest.

BLM manual direction and handbook guidelines govern how the BLM conducts mineral examinations and prepares mineral reports. Present Forest Service practice is to follow the procedures and guidelines established by the BLM. The BLM *Handbook for Mineral Examiners* was developed cooperatively between the BLM and the Forest

Service and provides guidance to government mineral examiners.

The Intermountain Region of the Forest Service which includes both Nevada national forests and 14 other national forests in Utah, Idaho, and Wyoming, has conducted approximately 52 mineral examinations over the past 4 years. In every case, the claimant has been notified by certified letter prior to the examination and asked to be present and participate in the examination. In most cases the claimant or his representative was present.

The 1983 case referred to in the article wherein the claimant had not received notification of the impending mineral examination is an isolated case involving an unfortunate set of circumstances. In this instance, the Forest Service notified the claim operator who was believed to be a representative of the claimant. The operator was present and involved during the mineral examination. It was not until years later that the claimant charged that the operator was not his official representative.

The Forest Service takes its responsibility for administering mineral activities very seriously. Our mineral examiners are well-trained professionals. Many are registered professional geologists in various states and belong to a variety of professional engineering and geological organizations such as the American Institute of Professional Geologists and American Institute of Mining Engineers. The professionalism of Forest Service mineral examiners is well recognized by the BLM and other government agencies.

Thank you for letting us respond to the article.

E.R. Browning
Director, Minerals Area Management

Letter from Denton W. Carlson:

I would like to respond to Mr. Dave Parkhurst's article in the December edition of the CMJ. It is a shame that Mr. Parkhurst judges the entire Forest Service on the basis of a singular Nevada case. Normally, I would not respond since the case involved is not a California case. But, since the Pacific Southwest Region of the Forest Service is included by implication, I am concerned.

Mr. Parkhurst's article, "USFS Mineral Examinations Violate Claimant's Rights," alludes the Forest Service is not professional, avoids appropriate procedures, and tramples on claimants' rights to "due process," and all this is based on one case which he refuses to identify. His allegations regarding the Forest Service cannot be farther from the truth, as can be attested to by many small-scale miners throughout the State of California.

Because relations between the Forest Service and the small-scale mining community have not always been amicable, we, along with some small-scale miners, formed the Regional Ad Hoc Committee of Forest Service and Small-Scale Miners in early 1984. The Committee's objective was to improve relations and increase the understanding of mutual problems between the two groups. The Committee has now been in existence for almost four years. The goals are to develop a forum where problems and concerns can be openly and freely discussed between the Forest Service and representatives of the small-scale mining community.

Membership consists of concerned Forest personnel and small-scale miner representatives, mainly from the Western Mining Council and Mother Lode Miners. The Committee is chaired by Tom Payne, president of Western Mining Council, Inc.'s Los Angeles chapter.

The Regional Committee has served as a model for Forests throughout the Region. As a result of the Regional example, many forests have formed similar committees, meeting on a regular basis with the local miners to address local problems.

One of the first projects of the Regional Committee was to develop a brochure or handbook on mining on National Forests. That project was accomplished in September, 1987, with the publication by the Western Mining Council of the mutually developed pamphlet "Mining on the National Forests in California." The pamphlet has been published by the CMJ and is available from the Western Mining Council and from Forest Service offices in California.

Much more than the development of the pamphlet was accomplished. The pamphlet's development served as the vehicle for many hours of discussion and agreements which resulted in achieving an understanding of, and a mutual respect for, the problems and concerns of the Forest Service and small-scale miners.

Since the formation of the Committee, there are many examples of where small-scale miners and Forest Service personnel have worked together harmoniously to solve mutual problems which has eliminated much of the negative press directed against the Forest Service in connection with small-scale miners' matters.

Many people in Region 5 have contributed to this turnaround in our relationship with the small-scale mining industry. Much credit goes to the small-scale miners whose initiative and willingness to meet with Forest Service personnel prompted the turnaround in relations. Credit also goes to the Forest Supervisors and their staffs for their meetings with the miners at the local level.

We still have our disagreements, but we are able to sit down and talk about them and, at least, understand each other's positions and the reason for it.

Thank you for letting me express our concerns towards Mr. Parkhurst's article and some of the recent history regarding the Pacific Southwest Region's Minerals Area Management policy.

Denton W. Carlson
Assistant Regional Forester
For Minerals Area Management

CMJ Editor's Note in January 1988 issue: In view of the Forest Service "Letters to the Editor" in response to last month's article "USFS Mineral Examinations Violate Claimant's Rights," it is obvious that the writers considered the article a pervasive criticism of F.S. mineral management activities. This was not the intent as we certainly realize and acknowledge the great strides that have taken place in Forest Service/Small-Scale Miner relationships as well as expanded mineral interest and expertise.

The reason we decided to publish this article is that it represents one complaint out of many that are sent to CMJ in letter form where we were able to obtain both sides of the story—the claimant's and the Forest Service's. Its purpose was to point out that F.S. policies and regulations are often not consistent and dependable, and that miners should continue to be alert to protect their interests.

The fact that three regional Forest Service Supervisors, two U.S.F.S. mineral examiners, and representatives in the U.S.F.S. legal department all told the article's author that the U.S.F.S. is **not required** to notify mining claimants of pending validity examinations, but that the procedure is normally followed—shows that at the very least the people involved in the case in point apparently weren't aware of any **requirement** to follow the BLM *Handbook of Mineral Examiners* or **requirement** to notify the claimant of pending validity examination.

The Regional Ad Hoc Committee of F.S. and Small-Scale Miners started in 1984 in California has, as Denton Carlson, Assistant Regional Forester for Minerals Area Management for the Pacific Southwest Region indicated, made great strides in F.S./ Small-Scale Miner relationships, understanding and problem solving. I am a member of this group and have witnessed this first hand. We would wholeheartedly encourage all F.S. regions to adopt this concept of F.S./Small-Scale Miner problem resolution.

But, at the same time, we must recognize that the catalyst for positive activity such as this is being constantly vigilant to protect our small mining interests. We feel this can only be accomplished by keeping informed on what the problems of miners are—this was the purpose of the CMJ article and an important responsibility of CMJ as a mining trade publication.

Letter from Don Robinson:

The December article by Dave Parkhurst on "USFS Mineral Examinations Violate Claimant's Rights" raises a serious concern about Forest Service policy on mineral examinations. I share the concern that Mr. Parkhurst has.

As a founding member of the Small Miners-Forest Service Committee for the Pacific Southwest Region, we are seeking some definitions and understanding of Forest Service regulations. We have outlined four areas for our upcoming meetings. Mineral Validity testing is one of the four. The Forest Service needs to define their policies, and implement them uniformly throughout our country. Their definition and the mining laws need to be consistent and non-conflicting.

It is our job representing the small miner at this committee to strive for this consistency and uniformity with the law. Along this line, it would be good for Mr. Parkhurst to attend and participate in our meeting on claim validity.

In fairness to the Pacific Southwest Region of the Forest Service, I will say they have made large strides towards changing their image and policies with respect to the miner. They are working with us to bring about some changes that are essential.

This Small Miners-Forest Service Committee has made a good start, but it is a very long road.

                   Don Robinson
                   President, Mother Lode Miners

13
Author's Response to USFS Letters on Claimant's Rights Issue

Author's Note: This item is a follow-up to my original article (entitled "USFS Mineral Examinations Violate Claimant's Rights") that appeared in the December 1987 issue of the *CMJ* as well as a response to the USFS replies concerning that article (two letters to the editor) in the January 1988 *CMJ*.

BEFORE ADDRESSING THE ISSUES outlined in my original article, I would like to clarify a few points: (1) as stated by Mr. Kenneth Harn (Ed. Note following USFS letters—Jan. 1988 *CMJ)*, we are fully aware of the U.S. Forest Service's program to resolve potential conflicts with mining interests and to improve relations between the Forest Service and miners, and we wholeheartedly support these efforts; (2) I personally have no quarrel with most of the Forest Service's management policies or their personnel; and (3) I have worked closely with USFS personnel on many occasions and have found them to be very professional and cooperative in most instances.

I *do*, however, strongly object to those instances where USFS policies have been applied in an unfair and unreasonable manner, and to the individual agency personnel who feel they can follow the various regulations and policies "at their own discretion." This is the reason the article was written—to point out a problem area that needs to be corrected. This problem has not yet been "fixed."

When I became aware of the inconsistencies evident in the conduct of certain USFS mineral examinations (several, in fact), I contacted a number of agency personnel (regional supervisors, mineral examiners, district rangers and others) and asked them this question: "Is the Forest Service required to notify a mining claimant prior to the initiation of a mineral examination on their properties?" They *all* said *no!* They all, also, added that this procedure was "normally followed." I then asked them if they were required to follow the BLM mineral examination procedures. They *all* said *no!* They explained that the various mineral examination procedures and handbooks were merely "guidelines," and that the Forest Service was not required to follow them.

Mind you, I had been told the same thing in a hearing before an

Administrative Law Judge in Reno on *October 19-20, 1987*. A mineral examiner and a geologist with the USFS both testified under oath that the Forest Service was *not* required to notify a mining claimant prior to the conduct of a mineral examination on their claims. They specifically said that the BLM mineral examination procedures were merely "guidelines" and that the Forest Service was not bound by them.

Since I found this difficult to believe, I obtained exact copies of the BLM 3891 regulations (Validity Examinations) and the U.S. Forest Service Title 2819 regulations (Mining Claim Contests). After carefully reading both sets of regulations, I personally visited two USFS district offices and asked for additional rules and/or regulations related to the conduct of mineral examinations. The agency personnel were very helpful, searched through their regulations and manuals, and informed me that I already had all the relevant materials. Needless to say, I had found that the Forest Service was correct in stating that they were not required to notify the mining claimant and that they did not have to follow the BLM mineral examination procedures. I then followed up with the telephone calls to USFS regional offices and other departments (late October and early November *1987*). As mentioned above, I got the same answers. Since the problem area had now been positively identified, I then proceeded to write the article.

As I mentioned in the article, I purposely avoided identifying the particular case and the individuals involved because I did not wish to prejudice the case and/or create any problems for the people involved, whether it be U.S. Forest Service personnel or the mining claimant. I have *not* refused to identify the case or individuals at any time, and have, in fact, discussed the exact names, places and times with a Forest Service legal representative (Mr. Dave Young, in Reno). No one from the USFS has contacted me at any time for details concerning either the case or the article despite the fact that I have made it known that I am available at any time. In contrast, I have made 48 telephone calls to USFS personnel and visited USFS offices 8 times since this issue came up last October.

Now, let us move on to the USFS letters sent in response to the article. Mr. E.R. Browning's letter would lead one to believe that the U.S. Forest Service consistently follows the mineral examination procedures and guidelines established by the BLM. Why, then, in October 1987, did a USFS mineral examiner and a geologist from a regional office flatly state that this is not required? Why would two regional Forest Service supervisors say the same thing in early November 1987? If the 1957 Memorandum of Agreement between the Forest Service and the BLM "establishes the work procedures governing action on claims for lands within National Forests," then why were these procedures not followed in 1983, 1984 and 1985, and why would the USFS personnel say they were not "bound" by these regulations in the fall of 1987? There doesn't seem to be much consistency here.

Mr. Browning states that the claim operator (an excavating contractor) was believed to be a representative of the claimant, and this operator was present and involved during the mineral examination. Why, then, wasn't this operator ever told that a mineral examination was in progress (verified by testimony under oath in October 1987)? Why was no attempt made to notify the mining claimant himself, especially since he had posted the required bond with the Forest Service and was specifically listed as the mining operator on the mining plan of operations? Why did the USFS refuse to request or consider any of the previous geological and mineralogical information compiled on the properties involved?

In addition, Browning asserts that "it was not until years later that the claimant charged that the operator was not his official representative." How could the claimant have done otherwise when neither he nor the "operator" was ever notified that a mineral examination was in progress until, well after the fact, a contest action was initiated in *May 1986*? And, how could the claimant respond in any manner when he did not receive a copy of the mineral examiner's report (his first opportunity to become aware of what happened in the case) until *late February 1987*? There seem to be a few discrepancies here also. If it is the wish of Mr. Browning and/or the U.S. Forest Service, I would be perfectly willing to publish an *exact* account of all the details involved in the case—including all of the names, dates, places and events arranged in chronological order. Almost all of the relevant information is included in a transcript of the October 1987 hearing, as well as from numerous other sources, and the information can be substantiated by the testimony and statements made by at least a dozen credible witnesses. But, is this necessary? Why not just correct the problem?

Mr. Denton W. Carlson's letter says it is a shame that I judged the entire Forest Service on the basis of a singular Nevada case. First off, I did not judge the Forest Service in any way. I stated *exactly* what I was *told* by the U.S. Forest Service—both by government witnesses under oath and by management personnel in telephone conversations. I also stated the *exact* sequence of events in the particular case involved. I'm not attempting to judge the individuals involved, either. I *am* pointing out a problem: the U.S. Forest Service should be *required,* by the regulation, to notify the mining claimant prior to conducting a mineral examination, and they should be *required* to follow a consistent, reliable mineral examination process. This has nothing to do with an individual case, unique or not. It concerns a lack of definitive regulations for the specific issues involved on a nationwide basis, as well as a policy that allows individual personnel to apply Forest Service policies on a "discretionary" basis.

Mr. Carlson also said that I have refused to identify the case and that my "allegations regarding the Forest Service cannot be farther from the truth, as can be attested to by many small-scale miners throughout the State of California." As stated above, I have not refused to identify the case (or any of the relevant particulars) to anyone who has asked me for this information, for the reasons noted. If we would like to get down to the nuts and bolts of the matter, I would be perfectly willing to do so.

The statement that my allegations cannot be farther from the truth would imply that I either do not have my facts straight or that I am, in fact, lying about the issue. If this is the case, why was it that neither letter specifically denied any of the factual information in the article? It seems to me that *both* letters skirted the actual issue involved, as neither letter mentioned specific Forest Service regulations that adequately cover mineral examination procedures. Mr. Browning said that present Forest Service *practice* is "to follow the procedures and guidelines established by the BLM" and that the work procedures governing action on claims by the USFS was established by the 1957 Memorandum of Agreement. Why, then, didn't Browning or Carlson state that the Forest Service is *required* to follow these regulations, or that this deficiency would be corrected?

I have great respect for the U.S. Forest Service, Mr. Carlson, Mr. Browning and other USFS personnel, and I would be more than willing to work with their personnel in resolving the problem.

14
USFS Proposal Would Bypass Mining Laws

HERE WE GO AGAIN! The U.S. Forest Service is now proposing a change in its regulations that would allow the agency to exempt itself from certain provisions of the mining laws that cover "common" and "uncommon" varieties of minerals. If approved as proposed, the Forest Service and the Bureau of Land Management would be operating under entirely different rules in regard to the determination of just exactly which minerals are salable and leasable from the federal government and those that are "locatable" under the general provisions of the mining laws, as amended.

In effect, the new USFS regulations would eliminate a large number of minerals and *uses of many other minerals* from the basic definition of the so-called uncommon and locatable minerals under current mining laws and acts of Congress. The regulations would allow the Forest Service to ignore the definitions of common and uncommon mineral varieties specified in the Act of July 23, 1955 (69 Stat. 367, chapter 375, section 3) and under 30 U.S.C. 611 of the mining laws, which reads as follows:

> No deposit of common varieties of sand, stone, gravel, pumice, pumicite, or cinders and no deposit of petrified wood shall be deemed a valuable mineral deposit within the meaning of the mining laws of the United States so as to give effective validity to any mining claim hereafter located under such mining laws: **Provided, however,** that nothing herein shall affect the validity of any mining location based upon the discovery of some other mineral occurring in or in association with such a deposit. **"Common varieties" as used in this subchapter and sections 601 and 603 of this title does not include deposits of such materials which are valuable because the deposit has some property giving it distinct and special value** and does not include so-called "block pumice" which occurs in nature in pieces having one dimension of two inches or more. "Petrified Wood" as used in this subchapter and sections 601 and 603 of this title means agatized, opalized, petrified, or silicified wood, or any material formed by the replacement of wood by silica or other matter.

The proposed USFS rule change was published in the Federal Register, Vol. 53, No. 82, on April 28, 1988, and all comments on the proposal had to be received by June 27, 1988. No public hearings were held on the issue, largely because the Forest Service also exempted itself from this requirement by citing the following reasons:

> This rule has been reviewed under Executive Order 12291 and USDA procedures and it has been determined that this rule is not a major rule. The rule would not have an economic effect of $100 million or more or affect U.S. competition in foreign markets. Additionally, it will not have a significant economic effect on a substantial number of small entities as defined under the Regulatory Flexibility Act (5 U.S.C. 601 et seq.). The proposed rulemaking contains no information collection requirements needing the approval of the Office of Management and Budget under 44 U.S.C. 3501.

This is a rather interesting interpretation by the Forest Service, since the rule change would allow the agency to substitute its own regulations in lieu of following specific directions contained in several acts by Congress, and the new regulations would directly impact the conduct of numerous mining operations and thousands of miners. The proposal would also create separate standards for determining common and uncommon varieties of minerals by the U.S. Department of Agriculture and the U.S. Department of the Interior. This is not a *major* change in the regulations?

The proposed new categories and *representative uses* within each category as delineated by the USFS are listed below (emphasis added by author).

Common Varieties

    A. *Agricultural Supply and Animal Husbandry Materials.* This category includes, *but is not limited to,* materials for: soil conditioners or amendments, fertilizers or other direct applications to the soil such as carbonate rocks, animal feed supplements, and other animal care products.

    B. *Building Materials:* Except for materials identified as Uncommon Varieties, this category includes, *but is not limited to,* materials such as: flagstone, ashlar, rubble, mortar, brick, tile, and terrazzo used for nonstructural components in floors, walls, roofs, fireplaces, and similar building construction uses.

    C. *Cleaning and Abrasive Materials:* This category includes, *but is not limited to,* materials used as or for: filters, absorbents, filing, scouring, polishing, sanding, and sandblasting.

    D. *Construction Materials:* This category includes, *but is not limited to,* materials used as or for: fill, borrow, riprap, ballast, road base or surfacing, crushed rock, concrete aggregate, and clay sealants.

E. *Decorative and Ornamental Arts Materials:* This category includes, *but is not limited to,* materials used as or for: sculpture, lapidary, furniture, and natural art objects. It does not include *precious gems.*

F. *Landscaping Materials:* This category includes, *but is not limited to,* chips, granules, sand, pebbles, scoria, cinders, cobbles, boulders, or slabs used for retaining walls, walkways, patios, yards, gardens, and the like.

Uncommon Varieties

A. Limestone suitable *and used,* without substantial admixture, for cement manufacture, metallurgy, production of quicklime, sugar refining, whiting, fillers, paper manufacturer, and desulfurization of stack gases.

B. Silica suitable *and used* for glass manufacture, production of metallic silicon, flux, and rock wool.

C. Alumino-silicates or clays suitable *and used* for production of aluminum, ceramics, drilling mud, taconite binder, foundry castings and other specific uses *for which there are no substitutes.*

D. Gypsum suitable *and used* for wallboard, plaster, or cement.

E. Block pumice which occurs in nature in pieces having one dimension of two inches or more.

F. Stone *recognized through marketing factors* for its special and distinct properties of strength and durability making it suitable for structural support and *used for that purpose.*

On the surface, it would not seem that the proposed change in USFS regulations would have a large impact upon the minerals industry. However, please note that the definitions under common varieties have *no limits* as to which minerals can be classed as being salable and leasable from the federal government, the definitions falling under uncommon varieties *are specifically restricted to particular uses,* and a strict interpretation of the new regulations would remove many currently locatable minerals and certain uses for other minerals from the purview of the mining laws.

All semi-precious stones (including opal, agate, turquoise, jade, etc.) would become salable and leasable minerals under the new regulations. In effect, all rock hound and lapidary enthusiasts would be barred from taking specimens from the national forests because these materials would *belong* to the federal government. The same thing would apply to other decorative and ornamental minerals.

Gypsum could *only* be used for specified purposes, and any gypsum used for agricultural purposes would, technically, have to be either purchased or leased from the government. In other words, gypsum would be classed as *both* common and uncommon according to its specific end use.

Barite would be removed from under the purview of the mining laws, since it is most commonly used as a weighting agent in drilling muds by the oil drilling industry. Considering the plight of the barite industry due to cheap foreign imports, the additional costs associated with buying or leasing the mineral from the government would effectively eliminate the industry in this country if this same rule were applied by the BLM.

In fact, a paradox would be created where a mineral deposit located in a national forest could be classed as a common variety whereas the same mineral deposit, if located on BLM-administered lands, could be classed as an uncommon variety. In other words, virtually identical minerals could be located with mining claims on BLM lands but would have to be purchased or leased from the government if they occur on Forest Service lands. Under these regulations, mining claims could be invalidated on Forest Service lands where the same claims on the same mineral deposit if located on BLM lands would be perfectly valid.

Could this be another effort by the Forest Service to invalidate selected mining claims and eliminate certain types of mining operations on National Forest lands? How would the USFS treat a gypsum mining operation on Forest Service lands if half of the material was used for an approved end use (e.g., wallboard manufacture) and half of the material was being marketed for an unapproved use (e.g., agricultural applications)? What would Forest Service personnel do if they caught a rock hound removing several pounds of agate from lands under their jurisdiction? What would the USFS do about preexisting mining claims in the national forests that were located on uncommon variety minerals under the current definition which suddenly became common variety minerals under the new definition? And, what about the uncommon variety minerals that are currently being marketed for an unapproved use under the new regulations?

This is only the tip of the iceberg insofar as the probable and potential problems that would be created by these new regulations go. Consider that, as unusual as it sounds, industrial diamonds would be classed as a common variety of mineral. All decorative or ornamental stone that is not specifically used for structural support would become a common variety mineral, and the cost of purchasing or leasing this material would drive up building costs and give a competitive edge to foreign suppliers. Careful thought as to the overall economic impact of these regulations must not have been given by the USFS personnel who drafted them. Or was it?

The manner in which the regulations have been constructed and the specific *wording* used make them appear to have been carefully designed by environmental activists, such as the Sierra Club. Whether specifically designed for this purpose or not, the new regulations effectively empower the U.S. Forest Service to ignore legislation enacted by Congress. They are, in effect, an attempt to make an end run around the mining laws, thus enabling the USFS to bypass the requirements of the mining laws and the various amendments thereto. As such, it is an obvious attempt to legislate mining matters through the implementation of administrative regulations.

If the USFS wishes to clarify troublesome areas in the definitions of common and uncommon variety minerals, then the agency should approach Congress with suggestions for appropriate legislation to resolve the issue. In the meantime, all miners should protest the proposal to change the Forest Service regulations and the manner in which the proposal was advanced. *Any* significant proposed change in the mining regulations by *any* administrative agency should first be discussed in public hearings in order to assess the overall impact of the proposal.

It should also be kept in mind that the administration of mining and mineral resources falls under the jurisdiction of the U.S. Department of the Interior—*not* the U.S. Department of Agriculture or the U.S. Forest Service.

PART III

The Small-scale Miner as Sacrificial Lamb

"The only thing the measure will do effectively is accelerate the process of eliminating miners and mining from the nation's public lands. It therefore follows that this end result is politically acceptable to the majority of our elected officials in both the federal administration and the U.S. Congress. If we don't all get involved and do something about this now, we will be sacrificed on the altar of utopian environmental extremism."

15
Nevada Governor Opposes $100 Mining Claim 'Holding Fee'

THE FEDERAL OFFICE OF MANAGEMENT and Budget (OMB) has included a proposed annual mining claim rental fee (holding fee) on all unpatented claims as part of the fiscal year 1991 Department of the Interior budget. Originally considered as a $5 per acre fee, it has been changed to $100 per claim per year. The budget proposal states the fees must be paid in cash each year as a replacement for the current annual assessment work requirement. A fee increase of this magnitude would devastate many small miners and prospectors as well as U.S. mineral exploration overall.

Realizing that the proposed fee would have a major negative impact on the state's mineral exploration and development, Nevada Department of Minerals Director Russ Fields requested Governor Bob Miller to become involved in the issue. As a result, Gov. Miller prepared a letter to each member of Nevada's congressional delegation that states his firm opposition to the fee proposal. The text of the letter is as follows:

> I have recently learned that the Office of Management and Budget has proposed an annual fee on mining claims and removal of the 1872 Mining Law requirement to perform annual assessment work on mining laws. This proposal is included in the Administration's budget before Congress at this time.
>
> As you know, a viable hardrock mining industry is of extreme importance to Nevada. There are approximately 400,000 active mining claims in Nevada, nearly 2.5 times as many as the next most active state in the Nation, California. Nevada mines produce more gold and silver than any other state. Our industry also leads the Nation in the production of several other mineral commodities. In addition to its importance to the economy of Nevada and other Western states, domestic mining plays a critical role in reducing the need to import mineral products from foreign sources.
>
> Imposition of an annual rental fee will have the immediate effect of cancellation of many current and future exploration projects by both small and large interests. Once existing ore reserves are depleted, hardrock mining on Federal lands will decrease significantly because of reduced exploration efforts to identify new deposits. The removal of

the assessment work requirement retards the diligent development of minerals and enhances the opportunity for nonresident speculators to abuse the Mining Law.

As a revenue mechanism, a rental fee on mining claims will not produce significant new dollars. The current number of active mining claims can be expected to fall tremendously with the imposition of a new fee. Mining interests unwilling to pay the fee will simply let their claims lapse.

Fewer claims will obviously produce lower total revenues. In the longer term, other revenues derived by the states and the Federal government from hardrock mining will fall because of the reduction in activity. Finally, the cost to the Nation in terms of paying for imported mineral products and the loss of domestic production of strategic minerals will be substantial.

An evaluation of the total impacts of increasing the cost of holding and developing unpatented mining claims under the 1872 Mining Law strongly suggests that the proposal made in OMB's FY1991 budget document will benefit neither the Nation nor the hardrock mineral-producing states. I ask that you urge Congress to withhold any support of the proposal to impose a rental fee on mining claims and removal of the requirement to perform annual assessment work.

(Signed: Nevada Governor Bob Miller)

Miners are concerned that this kind of supposedly revenue-raising fee would be particularly attractive to certain politicians, antimining groups and environmental extremists—especially in what is not only an election year but a year that includes a celebration of the 20th anniversary of Earth Day. They are being urged to organize a massive grassroots movement wherein all small miners, prospectors and mining companies will write letters and telephone their respective Congressional delegates, informing them of our individual and collective opposition to this fee. Many mining associations and organizations are also preparing and delivering factual testimony on the negative impacts of this proposal directly to congressional committees and individual members of Congress.

16
$100 Mining Claim Fee Proposal Defeated

THE NATION'S MINERS WERE again caught off guard by renewed federal efforts to impose a $100 fee per mining claim per year as a substitute for the current annual assessment work requirement. The fee was originally proposed by the Office of Management and Budget (OMB) last January as a part of the FY1991 Department of the Interior budget appropriations but, through the efforts of several Congressmen and the mining industry, the fee proposal was effectively killed in the U.S. House of Representatives.

However, the $100 mining claim fee proposal was reinserted in the budget summit negotiations by the OMB and the federal administration. The "deficit reduction package" sent to Congress contained language similar to that originally proposed by the OMB, but the budget summit agreement was defeated in the House on October 4. Prior to the House vote, mining representatives had mounted a last-minute campaign to contact all senators and representatives in Congress and inform them of the negative impact the fee would have on mineral exploration in the U.S.

The subsequent negotiated compromise in the deficit reduction resolution allowed for changes in the specific areas of budget cuts and tax and fee increases, so each budget item was then determined by the Senate and House committees. Consequently, the House Interior and Insular Affairs Committee did not include the fee in its list of plans to meet the deficit reduction requirements under the budget compromise.

The Senate Energy and Natural Resources Subcommittee took action on the $100 fee proposal, however, and the measure failed on a tie vote of 9 to 9. The committee reopened discussions of the budget reconciliation process the following week (on October 15) but the fee proposal was not reintroduced. Therefore, neither the Senate Energy nor House Interior Committee budget packages included the fee. As a result, the final budget package passed by both the House and the Senate did not include the mining claim fee proposal, and the measure was effectively defeated this year. However, it is expected that a similar proposal will resurface again in the next session of Congress.

It is interesting to note that an analysis of the projected mining claim holding fee revenues in the budget summit agreement, as shown in Table II.B, "Mandatory/Entitlements/User Fees—OMB and CBO Scoring," reveals major discrepancies in both the projected revenues and the overall impact on the mining industry. The Congressional Budget Office (CBO) had estimated total revenues of $120 million over the next five years from the fee, which means they believed the number of existing claims (about 1.2 million) would drop to an average of 240,000 claims over the 5-year period—for an 80% net drop from current levels.

However, the OMB only projected a drop of 420,000 claims from current levels, estimated an average of 780,000 claims over the 5-year period, and projected revenues of $390 million over the 5 years—based upon the assumption that only 35% of the current mining claims would be dropped. The mining industry has estimated that over 70% of existing mining claims would be dropped if the hardrock mining claim holding fee was approved, which shows that the estimate by the CBO is much more realistic.

It appears that the federal administration, and certain congressmen, were quite willing to sacrifice mineral exploration and development in the United States in order to obtain a fairly short-term increase in federal tax revenues. Therefore, miners will have to keep a close eye on the administration and Congress next year to ensure that this type of fee is not enacted into law.

17
OMB Resurrects $100 Mining Claim 'Holding Fee'

FOR THE THIRD TIME DURING the last year, the federal administration's Office of Management and Budget has proposed an annual $100 mining claim "holding fee" as a replacement for the current annual assessment work requirement under the mining law. This year, however, the OMB FY1992 budget appropriations language is much more detailed, and the Act specifically amends several sections of the 1872 Mining Law and the Federal Land Policy and Management Act of 1976 (FLPMA).

A careful analysis of the fee proposal reveals that it is much more than a simple tradeoff of cash payments in lieu of performing assessment work. First, however, it should be emphasized that the cash payment is not optional (there is no choice between paying the fee and doing the work), and if the fee is not paid each year then mining claims will automatically be declared null and void. The other impacts of the fee proposal can best be explained by providing the exact language in the Department of the Interior and Related Agencies Appropriations Act, 1991, followed by a brief analysis of the meaning and intent of each statement and, finally, a summary of the major impacts of the proposal.

The Appropriations Act language is as follows:
OMB Mining Claim Holding Fee Proposal
1. "...Provided further, that notwithstanding any other provision of law, that effective upon the date of enactment of this Act for the fiscal year 1992 and every year thereafter, for each unpatented mining claim, mill or tunnel site on federally owned lands, in lieu of the assessment work requirements contained in the Mining Law of 1812 (30 U.S.C.28-28[e]), and the filing requirements contained in Section 314(a) of the Federal Land Policy and Management Act of 1976 (FLPMA) (43 U.S.C.1744[a]) and the related requirements of Section 314(c) of FLPMA (43 U.S.C.1744[c]), the claimant shall pay an annual holding fee of $100 to the Secretary of the Interior or his designee on or before August 31 of each year in order for the claimant to hold such unpatented mining claim, mill or tunnel site for the following year beginning on September 1;"

Meaning and Intent: Beginning in FY1992 (after Oct. 1, 1991) and thereafter, claimants pay an annual $100 fee in cash instead of performing the federal annual Proof of Labor with BLM; fee must be paid on or before August 31 of each year to hold mining claims for the next year starting on September 1; fee intended to eliminate land disturbances caused by assessment work; fee intended to produce relinquishment of speculative mining claims; fee revenues to cover BLM administrative costs; majority of fee revenues deposited in the U.S. Treasury as income; *state annual assessment work requirements are not affected by the federal holding fee;* and the *administration will urge the states to remove assessment work requirements and collect annual fees to hold claims under state laws.* In other words, miners will have to perform state annual assessment work *and* pay the $100 annual federal fee, or they will have to pay a state annual holding fee *and* pay the $100 annual federal fee.

2. "...Provided further, that the fee established by this Act in lieu of the assessment work requirements for the assessment year ending at noon on September 1, 1992, shall be due and payable to the Secretary on or before June 30, 1992, except that such fee otherwise due and payable for the period shall be waived by the Secretary or his designee if the claimant files an affidavit of assessment work by June 30, 1992, showing the labor required by 30 U.S.C.28 was completed for the assessment year ending at noon September 1, 1992, before the effective date of this Act; provided further, that such fee otherwise due and payable for the assessment year ending at noon on September 1, 1992, for mill and tunnel sites shall be waived by the Secretary or his designee if the claimant files a Notice of Intention to Hold the site by June 30, 1992;"

Meaning and Intent: Provides for transition from current assessment work requirement to annual holding fee; 1992 fee is waived if assessment work is completed prior to the effective date of this Act (when it is actually passed by Congress) and if affidavit is filed with BLM by June 30, 1992; and 1992 fee is waived if Notice of Intent to Hold mill and/or tunnel sites is filed by June 30, 1992.

3. "...Provided further, that for every unpatented mining claim, mill or tunnel site located after the date of enactment of this Act, the locator shall pay $100 to the Secretary of the Interior or his designee

at the time the location notice is recorded with the Bureau of Land Management to hold such claim for the year in which the location was made."

Meaning and Intent: When claimants record initial Location Notices or Certificates of Location with the BLM for new unpatented mining claims, mill or tunnel sites located after the enact of this Act, they must also pay the $100 holding fee in order to hold claims for the year following the date upon which the claims were recorded. In other words, *the initial recording costs for new mining claims increase from $10 to $110 per claim.*

4. "...Provided further, that the co-ownership provision of 30 U.S.C.28 will remain in effect except that the annual holding fee shall replace the assessment work requirements and expenditures;"

Meaning and Intent: Co-owners are still responsible for their share of claim maintenance costs under the General Mining Law.

5. "...Provided further, that failure to make the annual payment of the holding fee required by this Act shall conclusively constitute an abandonment of the unpatented mining claim, mill or tunnel site by the claimant."

Meaning and Intent: Failure to pay the annual holding fee on or before August 31 of each year results in an invalidation of mining claims (see exceptions for 1992).

6. "...Provided further, that nothing in this Act shall change or modify the requirements of Section 314(b) of FLPMA (43 U.S.C.1744 [b]) or the requirements of Section 314(c) of FLPMA (43 U.S.C.1744 [c]) related to filings required by Section 314(b), which shall remain in effect."

Meaning and Intent: The BLM requirement for filing and recording Certificates of Location for new mining claims remains in effect, and they are not modified in any way by this Act.

7. "...Provided further, that the Secretary of the Interior shall promulgate rules and regulations to carry out the purposes of this Section as soon as practicable after the effective date of this Act."

Meaning and Intent: Directs the Department of the Interior to formulate new regulations and amend existing regulations to comply with the provisions of this Act as soon as possible after enactment.

Overview and Analysis of $100 Fee Proposal

1. The mining claim holding fee is counterproductive: It would discourage active minerals exploration and development while it would encourage speculative enterprises by moneyed interests.

2. The fee would produce an onerous burden on small miners and prospectors, most of whom can afford the time to perform the annual assessment work requirements but cannot afford cash payments. By not preempting state and other federal fees and requirements, miners would probably be faced with the payment of a holding fee to both the federal government and the state (or still have to perform $100 in labor each year), and continue to pay property taxes, state filing fees, and state and federal agency permit fees.

3. The additional costs would produce a forced relinquishment of many mining claims by small miners, who it is estimated currently hold about 8% of the total claims in the U.S. Recent polls indicate that miners would have to drop at least 70% of their claims.

4. The OMB projects revenues from the mining claim holding fee at $400 million over a 5-year period ($80 million per year). Last year, the Congressional Budget Office (CBO) estimated the fee would produce revenues of $120 million over 5 years ($24 million per year). According to mining industry estimates, the CBO figure is much more realistic. This would indicate that the budget proposal would actually produce less than 30% of the OMB's projected revenues.

5. Over the longer term, the resultant decline in mineral exploration activity would produce a dramatic decline in the nation's mine production, because new mineral deposits would not be discovered and developed to replace those that are currently being mined out. This would, in turn, produce a major decline in federal and state revenues from minerals production and associated activities.

18
$100 Claim 'Holding Fee' Deleted from HR 2686

REP. BARBARA VUCANOVICH (R-NV) was successful in removing the administration's proposed $100 mining claim holding fee from H.R. 2686, which contains the Department of the Interior's Bureau of Land Management FY1992 budget appropriations. However, Rep. Vucanovich's effort to strike the moratorium in the same bill on the BLM's issuance of mining claim patents under the Mining Law was unsuccessful, and the patenting moratorium provision remains in the appropriations bill.

In her remarks on H.R. 2686 on June 25, 1991, Vucanovich stated the following:

> Mr. Speaker, this revenue estimate (CBO scoring of $40 million of revenue to the Federal Treasury) is without foundation. The CBO and OMB are simply guessing at the elastic response the mining community would have to this new tax. The proposal is that mining claimants simply send the money that they are now required to spend on development of their claims to Washington, D.C., instead.
>
> This [proposed $100 mining claim holding fee] will do nothing to aid in finding ore deposits or in cultivating mines on the public lands in the West where the Mining Law operates. In fact, it would devastate the economies of rural areas in Nevada, and elsewhere, which are dependent upon miners spending their exploration and development dollars locally.
>
> Chairman Rahall and I both agree that diligent development of the mineral lands in the public domain is in the Nation's best interest. Our Mining and Natural Resources Subcommittee is currently considering legislation (H.R.918) to reform the Mining Law of 1872, as amended. While we may disagree on the reform issue, we do agree that the Interior Committee should be given proper opportunity to act and that the appropriators ought not to legislate actions that would undo the deliberations of the authorizers.

As a direct result of her efforts, the $100 mining claim holding fee was deleted from the budget appropriations package contained in H.R. 2686.

Rep. Vucanovich also stated:

> For these same reasons, I have sought to strike the moratorium in this

bill on BLM's issuance of patents under the Mining Law. Unfortunately, Chairman Rahall does not agree with me on this issue, and the patenting moratorium will remain in the bill to be sent to the Senate.

The patent moratorium language in H.R. 2686 reads as follows:

"Provided further, that none of the funds appropriated or otherwise made available pursuant to this Act shall be obligated or expended to accept or process applications for a patent for any mining or mill site claim located under the general mining laws or to issue a patent for any mining or mill site claim located under the general mining laws unless the Secretary of the Interior determines that, for the claim concerned: (1) a patent application was filed with the Secretary on or before the date of enactment of this Act, and (2) all requirements established under sections 2325 and 2326 of the Revised Statutes (30 U.S.C. 29 and 30) for vein or lode claims and sections 2329, 2330, 2331, and 2333 of the Revised Statutes (30 U.S.C. 35, 36 and 37) for placer claims, and section 2337 of the Revised Statutes (30 U.S.C. 42) for mill site claims, as the case may be, were fully complied with by that date."

Rep. Vucanovich also addressed a provision in the bill that would impose a much higher grazing fee on ranchers using the public lands. She said:

> Likewise, although the Rules Committee issued a rule allowing a point of order to rest against the 33⅓ per centum grazing fee increases in the Appropriations Committee reported bill, they also saw fit to make the amendment to be offered by Mr. Synar in order. This amendment, which would invoke a much higher grazing fee increase, also constitutes legislating in an appropriations bill.
>
> Furthermore, the authorizing committees of jurisdiction, Interior and Insular Affairs, and Agriculture, both are looking at the grazing issue. There simply is no justification for this appropriations bill to be a vehicle to legislate an outrageous fee increase without proper hearings and opportunity for debate.

However, the increased grazing fee provision remained intact in the proposed legislation.

Since this is still fairly early in the budget appropriations process, there will still be an opportunity for someone to re-propose the $100 mining claim holding fee on the Senate side. However, it is expected that the Senate will most certainly oppose the hike in grazing fees and, judging by the 1990 session, very likely oppose the inclusion of the patent moratorium. Unfortunately, it is also common practice for some politicians to tack on unrelated spending and revenue-raising measures during the last few hours of the budget appropriations "negotiations." Again, judging by the past few sessions of Congress, it seems likely that these issues will not be finally resolved until near the end of October.

19
$100 Mining Claim 'Fee' in 1992 Federal Budget

AS COULD BE EXPECTED, the federal administration's Office of Management and Budget (OMB) has again proposed an annual $100 mining claim "holding fee" as a replacement for the current federal annual assessment work under U.S. mining law. The proposal is included in the Department of the Interior and Related Agencies Appropriations Act, 1992, which is part of the total federal budget package that President Bush has directed Congress to address by March 20, 1992.

Although the inclination and ability of Congress to approve a federal budget "Resolution" by the March deadline is questionable, the budget process has definitely been accelerated this year. The U.S. economic recession is exerting a tremendous pressure on Congress and the administration to institute measures that will provide a significant boost to economic activity as quickly as possible. Because 1992 is also an election year, politicians will also be seeking an economic upturn well before voters cast their ballots in November.

So, there might be less time available to adequately address the $100 fee proposal. This measure continues to be an attractive means for Congress to potentially enhance federal revenues during a time of massive deficits and a faltering national economy. In spite of the fact that the devastating impacts of this so-called fee on mining have been pointed out time and time again, the proponents of this "cash only" measure have chosen to ignore the facts and continue to proceed with the fantasy that the fee will actually increase revenues to the U.S. Treasury. The mining claim fee proposal is, of course, of major concern to the nation's mining community, especially to small-scale miners and exploration firms. Therefore, the same battle to oppose this measure must be fought over and over again, but this time it must be done effectively so it doesn't show up again next year.

The budget appropriations language this year is very similar to that proposed last year (with one notable exception), and enactment "as is" would require specific amendments to several sections contained in the 1872 Mining Law and the Federal Land Policy and Management Act of 1976 (FLPMA). [As mentioned in the previous article,] a careful analysis of the proposed Act reveals the fact that it is much more than a simple tradeoff involving cash payments in lieu of performing annual assessment work. For example, payment of the federal annual fee does not remove the obligation to meet all legal requirements under various state laws, so state annual assessment work and/or filing and fee requirements must still be met.

It should also be emphasized here that the annual cash payments are not optional—there is no choice between paying the fees and performing the work—and if the fee is not paid on time every year then the mining claims will automatically be declared null and void. In addition, expenditures incurred for exploration and development work on the claims cannot be charged against the annual cash payment required. In other words, you pay $100 per claim per year in cash *plus* all other costs associated with developing the claim and meeting state requirements.

Now, let's examine the notable exception referred to above. Appendix One-583 of Department of the Interior budget appropriations, under "Administrative Provisions" (beginning on line 1 through line 7), reads, in part: "Appropriations for the Bureau of Land Management shall be available for...; up to ($25,000) $100,000 for payments, at the discretion of the Secretary, for information or evidence concerning violations of laws administered by the Bureau of Land Management; ..."

What does this mean? The Act specifically provides a payment of up to $100,000 to any person reporting violations of the BLM's administrative laws, rules and regulations (an increase of $75,000 per violation over the level approved last year), and it applies to all activities, under the Bureau's jurisdiction.

What is this—increased regulatory enforcement by enlisting bounty hunters? And who says that all of the people that might be attracted by this provision are going to be honest? Most importantly, this type of *reward* provision illustrates a rather pervasive penal approach in federal legislation and regulation.

An examination of the projected revenues expected to be generated by this mining claim "fee" for the General Fund of the U.S. Treasury (read: money to be used to fund federal programs *not* related to mining) reveals the same mindless numbers-crunching that creates record-breaking federal budget deficits every year. However, the projected revenues will be allocated to other programs and will be spent, meaning that this proposal will, in the final outcome, produce an increase in the federal deficit. The fallacies underlying the OMB's revenue estimates will be pointed out and analyzed following a complete description of the budget proposal.

The overall content and projected impacts of the fee proposal can best be explained by excerpting specific portions of the Act, followed by a brief analysis of each statement, and then providing an analytical summary of the major impacts that will be created by the proposal. The language below is excerpted directly from Appendix One-583, Administrative Provisions (beginning at line 36), of the Department of the Interior and Related Agencies Appropriations Act, 1992:

1) "...Provided further, That notwithstanding any other provisions of law, that effective upon the date of enactment of this Act for the fiscal year 1993 and every year thereafter, for each unpatented mining claim, mill or tunnel site on federally-owned lands, in lieu of the assessment work requirements contained in the Mining Law of 1872 (30 U.S.C. 28-28[e]), and the filing requirements contained in Section 314(a) of the Federal Land Policy and Management Act of 1976 (FLPMA) (43 U.S.C. 1744[a]) and the related requirements of Section 314(c) of FLPMA (43 U.S.C. 1744[c]), the claimant shall pay an annual holding fee of $100.00 to the Secretary of the Interior or his designee on or before August 31 of each year in order for the claimant to hold such unpatented mining claim, mill or tunnel site for the following year beginning on September 1;..."

Summary and Analysis: After October 1, 1992, and thereafter, claimholders required to pay an annual holding fee of $100.00 per claim in cash instead of performing federal annual assessment work; replaces the filing of an annual Proof of Labor with the BLM; fees must be paid on or before August 31 of each year (see exception for 1993 below) to hold mining claims for the next year beginning on September 1; fee is intended to eliminate the land disturbance caused by assessment work; fees are intended to encourage relinquishment of speculative mining claims; majority fee revenues (about 82%) deposited in General Fund of U.S. Treasury; remaining revenue to fund BLM's mining law administration programs (slightly less than 18%); state mining claim requirements are not affected, and the administration will urge states to adopt more claim fees in lieu of assessment work.

2) "...Provided further, That the fee established by this Act in lieu of the assessment work requirements for the assessment year ending at noon on September 1, 1993, shall be due and payable to the Secretary on or before June 30, 1993, except that such fee otherwise due and payable for this period shall be waived by the Secretary or his designee if the claimant files an affidavit of assessment work by June 30, 1993, showing the labor required by 30 U.S.C. 28 was completed for the assessment year ending at noon September 1, 1993, before the effective date of this Act; Provided further, That such fee otherwise due and payable for the assessment year ending at noon on September 1, 1993, for mill and tunnel sites shall be waived by the Secretary or his designee if the claimant files a notice of intention to hold the site by June 30, 1993;..."

Summary and Analysis: Provides for a potential transition period from the requirement for annual assessment work to the new annual holding fee; the 1993 fee is waived if assessment work has been completed prior to the date of enactment of this Act (when the legislation is actually passed by Congress and signed by the president) and if the affidavit is filed with the BLM by June 30, 1993; the 1993 fee for mill and tunnel sites is also waived if a Notice of Intent to Hold is filed by June 30, 1993; but only cash payments of fees will be accepted (no waivers allowed) if the Act is enacted by Congress prior to September 1, 1992, and all such fees must be paid by June 30, 1993 for that year. For 1994 and the years following, the fees must be paid on or before August 31.

3): "...Provided Further, That for every unpatented mining claim, mill or tunnel site located after the date of enactment of this Act, the locator shall pay $100.00 to the Secretary of the Interior or his designee at the time the location notice (certificate) is recorded with the Bureau of Land Management to hold such claim for the year in which the location was made. Provided further, That the co-ownership provision of 30 U.S.C. 28 will remain in effect except that the annual holding fee shall replace the assessment work requirements and expenditures;..."

Summary and Analysis: When claimants record Location Notices or Certificates of Location with the BLM for all new unpatented mining claims, mill or tunnel sites located after the date of enactment of this Act, they must also pay the $100 holding fee in order to hold the new claims for the year following the date upon which the claims were recorded. Under current law, the cost and time associated with staking a mining claim, preparing maps and documents, and recording are accepted in lieu of assessment work for the first year, but this would no longer be allowed. The $100 fee must be paid in addition to the $10 BLM recording fee at the time the claims are recorded. In addition, co-owners are still responsible for their share of claim maintenance costs, but this is replaced with cash payments only, and they are still responsible for their share of development costs.

4) "...Provided further, That failure to make the annual payment of the holding fee required by this Act shall conclusively constitute an abandonment of the unpatented mining claim, mill or tunnel site by the claimant; Provided further, That nothing in this Act shall change or modify the requirements of Section 314(b) of FLPMA (43 U.S.C. 1744[b]) or the requirements of Section 314(c) of FLPMA (43 U.S.C. 1744[c]) related to filings required by Section 314(b), which shall remain in effect;..."

Summary and Analysis: The failure to pay annual holding fees on or before August 31 of each year (see specific exceptions for 1993) will result in an invalidation of mining claims. The current BLM recording and fee requirements for filing Certificates of Location for new mining claims will remain in effect and will not be modified by this Act.

5) "...Provided further, That the Secretary of the Interior shall promulgate rules and regulations to carry out the purposes of this Section as soon as practicable after the effective date of this Act."

Summary and Analysis: Directs the Department of the Interior (and the BLM) to formulate new regulations and amend existing regulations to comply with the provisions of this Act as soon as possible after enactment. This would also include regulations for implementing the "reward" provision, described previously, for persons reporting violations.

Impact Analysis of $100 Fee Proposal

The Interior's accompanying budget information for BLM operations (entitled "Justification of Programs and Performance," on page BLM 4-21) states, in part:

> The fee would generate an estimated revenue of $97,600,000. Of this total, the 1993 budget provides that $80,200,000 will go to the General Fund of the Treasury; ... BLM would have a total of $17,430,000 from the proposed holding fees to conduct the mining law program and collect the holding fees.

At $100 a whack, this means the total $97,600,000 budget projection is based upon 976,000 active mining claims in 1993. On the previous page (BLM-4-20), the BLM estimates a total of 1.1 million mining claims nationwide, yet states the agency processed 860,000 annual filings and recorded 81,000 new claims in 1991. Correct me if I'm wrong, but the BLM 1991 claim figures total 941,000 claims—which is already 35,000 claims below the base figure used for the budget estimate and 159,000 claims below the total 1.1 million active claim estimate. Therefore, in order to achieve the full projected revenues for the budget in 1993, it appears that the government administration is planning on an increase in the total number of mining claims *after* the $100 mining claims holding fee is in effect. Who does the math for these people?

In fact, mining claim data for 1991 actually show a fairly sharp drop in the total number of active mining claims—at least in Nevada, which has almost 40% of all the mining claims in the United States. In 1991, the number of new mining claims filed in Nevada dropped about 26.6% from 1990 levels (from 35,078 to 25,764) and were down 69% from the 1988 filings (83,389 to 25,764), based on the federal fiscal year (October through September). The total number of active claims in the state (new claim filings plus existing active claims) had dropped 5.3% by June 30, 1991, from the same time in 1990 (down to 308,839 from 326,064), and the total filings during the last half of 1991 were down about 12% from the same period one year earlier (Nevada Department of Minerals data). This would strongly indicate that the OMB's revenue estimate is unrealistic at present—even before the approval of the $100 fee—and there is still about 1.3 years to go before June 10, 1993.

2. The imposition of a $100 mining claim fee would disproportionately fall on the shoulders of those least able to pay the fee, primarily the small-scale miners, small exploration firms, and individual prospectors. It has been recently estimated that small-scale miners hold around 85% of all mining claims, so the federal revenues are, in

fact, targeted to fall on the backs of the poorest members of the nation's mining community. Recent polls of small-scale miners and prospectors indicate that the imposition of a $100 cash-only fee per claim per year would force the relinquishment of about 70% of the claims held by small-scale individuals and firms. They are willing, and able, to perform assessment work, but they just don't have the financial resources available to meet the higher cash payments.

3. Mineral exploration is a costly and risky business, and very few of those who are involved in this type of venture are actually successful. Yet the mineral exploration business amounts to millions of dollars in economic activity and provides much-needed employment, especially in the smaller and more remote communities in the West. At the present time, mineral exploration activity has already declined to the lowest levels seen in the past decade. Significant increases in cost, like the imposition of the $100 fee, would effectively kill mineral exploration by the individual and leave the field to speculative moneyed interests and large firms that can afford the additional expense. So, here we are in the midst of a nationwide economic recession, and the government proposes to enact a burdensome fee that would reduce economic activity, create unemployment and practically demolish an important segment of the U.S. minerals sector.

4. Over the longer term, the resultant decline in mineral exploration and development activity would produce a dramatic decline in the nation's mineral production, because new mineral deposits would not be discovered and developed to replace those that are currently being mined out. This would, in turn, greatly reduce economic activity, produce higher unemployment, and bring about a major decline in federal and state revenues from minerals production and businesses that supply the mining industry.

Many more negative impacts could be cited but those listed above should be sufficient to illustrate the major flaws in this legislative proposal. Government should be *encouraging* private enterprise and economic activity rather than actively contributing to their decline.

20
U.S. Government Planning to Sacrifice Small-scale Miners

WELL, IT'S FINALLY REACHED a point where anti-mining factions have pressured our government into a position where politicians must choose between taking a firm stand in defense of our nation's miners and caving in to unreasonable demands by mining's opponents. And, apparently, political expediency has won. Behind-the-scenes negotiations are currently being conducted, without any input from small-scale miners or mineral exploration firms being taken into account or, for that matter, even being asked for, whereby our supposed friends in Congress and the federal administration will agree to actively support the OMB's $100 mining claim holding fee proposal.

That's right. At least some of our supporters in Congress who have consistently opposed the imposition of the $100 mining claim fee, citing the extensive damage this fee would wreak on U.S. mineral exploration and the nation's small-scale miners and prospectors, have now agreed to use the fee as a political trade-off—thereby placing us on the auction block in the totally irrational hope that opposition to mining will disappear. They apparently think that anti-mining activists will be satisfied if all of the "little guys" are put out of business.

There is no supposition or assumption involved here: two of our most staunch defenders in the U.S. Senate have personally stated (one orally and one in a letter to a constituent) that they were going to support the imposition of the $100 fee. Cash only. No option of either performing the annual assessment work or paying the fee. Case closed.

The *only* reason the two senators' names do not appear here is because they have not yet *officially* laid the issue on the table in Congress. But they most assuredly will, unless *all* small-scale miners, prospectors and mineral exploration firms realize just how serious this situation is and actively respond by making telephone calls and writing letters to their congressional representatives. And miners had better not assume anything this time around: If you remember just how close the votes have been on this issue in the last two years, then you know we just squeaked through by the skin of our teeth. Two more votes for the OMB fee proposal (with the consequent two less votes against) would probably be enough to pass the measure, and it seems likely that more politicians have agreed to this trade-off.

For those who might be unfamiliar with the administration's Office of Management and Budget (OMB) $100 mining claim holding fee proposal in the Department of the Interior and Related Agencies Appropriations Act, 1992, please refer to the article entitled "$100 Mining Claim 'Fee' in 1992 Federal Budget" on page 21 of the March 1992 issue of the *CMJ* [the article immediately preceding this one, article 19]. This article provides the exact language contained in the summaries and an analysis of each section, and an impact analysis of the fee proposal. For that matter, quite a few analyses of this issue have been prepared and delivered to Congress over the past few years, and all of them point out the unfairness and unrealistic revenue projections as well as the extremely negative impacts on the U.S. mining community if this measure is approved. So, members of Congress are fully aware of the consequences.

If a certain amount of disgust and anger with our elected officials is noted by those reading this item, please keep in mind that they are totally justified. The mining claim holding fee proposal is one of the most ill-conceived, poorly thought-out and blatantly irrational pieces of legislation considered by the U.S. Congress for quite some time. The measure intentionally singles out a specific segment of American business and methodically plans its destruction for no rational reason. If this statement on the issue seems overly strong, consider the following:

1. Imposition of a $100 cash-only mining claim holding fee would create a financial burden that would fall primarily on the shoulders of those people who are least able to pay the fee: mainly small-scale miners, individual prospectors, and small exploration firms. Before implementing the 1989 fee increases on mining claim filings (adding a $5 per claim fee for the filing of annual Affidavits of Labor for assessment work and raising the fee for recording new mining claims, notices of intent to hold, amended claim certificates, and filing applications for patent), the Bureau of Land Management conducted a study which showed that small-scale miners held about 85% of all mining claims (based on claim holdings of 20 claims or less).

So, this federal revenue-raising measure is, in fact, targeted to fall on the backs of the poorest members of the nation's mining community. Polls of small-scale miners and prospectors show the cash-only fee per claim per year would force the relinquishment of between 70% and 80% of the claims held by individuals and small companies. Two years ago, the Congressional Budget Office itself estimated an average drop in the total number of mining claims in the U.S. of around 80% over a 5-year period.

2. The OMB's rationale for imposing the fee (attached to the Department of the Interior's Budget Appropriations proposal) specifically states that it also intends to reduce the number of "speculative" mining claims. By the elimination of most small-scale miners and prospectors as claimants, one must assume that this is exactly what is meant by the administration's intent to get rid of mining speculators. If the individual can no longer afford to participate in mineral exploration and development, then the entire field will be left open to only those who have the financial capability to compete in the acquisition of this country's mineral wealth. This is the exact opposite of the OMB's rationale, because it would specifically encourage speculative investment in mining by moneyed interests.

3. Although the OMB proposal purports to substitute the $100 fee for annual assessment work with the objective of reducing the environmental impacts of mining activity, the primary thrust of the measure is, in fact, directed toward increasing federal revenues. The OMB estimates the fee proposal will generate $97,600,000, of which the 1993 budget stipulates $80,200,000 will be deposited in the General Fund of the Treasury. *This means that over 82% of the fees collected from miners are meant to pay for other government spending programs that are entirely unrelated to mining.*

This amounts to selective taxation of a specific segment of society for the exclusive benefit of others.

4. You pay annual cash payments to the federal government and get absolutely nothing in return. The lands covered by your mining claims will remain open to use for recreation, hiking, grazing, timber harvesting, public wood-cutting, Christmas tree removal, and other multiple-use activities. The mining claimant is still subject to all surface management regulations, use fees, permit fees and massive civil and criminal penalties if, heaven forbid, you happen to be found to

be in noncompliance with the overwhelming number of local, state and federal laws, rules and regulations. In many states, you may still be required to perform annual assessment work (the $100 fee *only* replaces federal assessment work requirements). In fact, many states will most likely impose an additional annual mining claim fee, because this is being actively encouraged by the U.S. Government.

5. The annual fees are counterproductive; they actively discourage the exploration and development of new mineral deposits. Grassroots mineral exploration in the U.S. has already declined to the lowest levels seen in the past few decades, and a major portion of this decline can be attributed directly to the massive increase in regulatory costs and requirements. Additional totally unproductive mining costs, when added to the existing expense and risk associated with mineral exploration, would produce severe negative impacts on the country's ability to develop its mineral resources, particularly at a time when miners have already been pushed to the limit by pervasive overregulation.

Over the long term, the accelerated decline in mineral exploration and development will produce a dramatic drop in U.S. minerals production, because no new mineral deposits will be discovered and developed to replace those that are currently being mined out.

6. Just a few years ago, the mineral exploration business was providing millions of dollars in economic activity and much-needed employment in many of the smaller and more remote communities in the West. The nationwide economic recession, lower mineral prices and ever-increasing mining costs have already produced a significant number of layoffs in the U.S. minerals industries, and most of these have occurred in the exploration field. So, at a time when U.S. mining activity has already undergone a significant decline, the federal government wants to impose additional burdensome fees and excessive regulations that would substantially reduce economic activity, create additional unemployment, and practically destroy an important segment of the American minerals sector.

7. The $100 mining claim fee proposal is a budget-buster; the projected revenues of over $97 million is totally unrealistic, the projected $80 million to fund other government programs will not be received and, because the projected revenues will be spent, the appropriations

package will increase the federal deficit. Reliable statistics indicate that the fee will, in fact, generate less than 30% of the projected OMB revenue figure. The total number of mining claims in the U.S. is already declining, yet the OMB numbers-crunchers actually predict an increase in the number of active claims and assumes this number will remain the same *after* the $100 fee is imposed.

8. The most notable (and unbelievable) thing to be said about all of the impact analyses provided above consists of the fact that all of this information has been presented to the federal Administration and the U.S. Congress two years ago, again last year, and again this year. In other words, our elected government officials have sufficient information to know exactly what they are planning to do to the nation's small-scale mining community. It also means they must know the legislative measure is critically flawed, is unworkable, will not generate the projected revenues, will create more unemployment and will act to reduce U.S. economic activity.

The only thing the measure *will* do effectively is accelerate the process of eliminating miners and mining from the nation's public lands. It therefore follows that this end result is politically acceptable to the majority of our elected officials in both the federal administration and the U.S. Congress. If we don't all get involved and do something about this *now*, we will be sacrificed on the altar of utopian environmental extremism.

Think about it. Then act! We need to let our government know, in no uncertain terms, that this legislation is totally unacceptable.

21
Small-scale Miners Get the Shaft

THE U.S. CONGRESS AND THE FEDERAL administration have apparently decided upon a plan of action to appease the radical anti-mining extremists: the U.S. Government will put small-scale miners, self-employed consultants and junior mineral exploration firms out of business to temporarily satisfy the ravenous appetite of the utopian environmentalists. It seems that the politicians think they can still maintain American minerals production as they willingly give the nation's small miners the shaft.

The Interior Appropriations Act for FY1993 has become the vehicle for eliminating small-scale miners and explorationists from the minerals exploration and development business—at least on public domain lands. It can now truly be called "The Small-scale Miner Unemployment Act of 1993." If these statements appear to be extreme, consider the following:

1. In the Senate Interior Appropriations bill, the language in the Office of Management and Budget's original proposal has been changed to require that the $100 mining claim holding fee be paid by all claimholders next August 31 (the option of filing assessment work for 1993 has been removed). In other words, if you performed your assessment work during September of this year, thinking that at least you could hold your claims for one more year without paying the fee, you're mistaken. Your work and your expense have gone for naught: you still have to pay the $100 fee next year.

2. The appropriations language now has another added surprise: *You now have to pay for the following year, in advance!* In other words, the Congress wants money so badly that you will have to pay $200 per claim next August 31 ($100 for the current year and $100 for the following fiscal year). The *only* exceptions to this requirement are if you are *producing* from 10 or fewer claims *in an integrated operating area* which has *less than 10 acres of unreclaimed surface disturbance from mining activity*.

3. In addition, claimants must pay a $100 fee up-front whenever they file a new claim with the Bureau of Land Management, at the time the certificate is filed. Now, this is just the fee to hold the claim

for the year in which it was located—the assessment year ending August 31. Suppose you locate a new claim in the spring of the year. When the 31st of August rolls around, you must pay another $100 to hold the claim for the next year in advance (no provision for pro-rating).

4. Here's the real clincher. The budget appropriations process now contains a provision which *requires a claimant to post a bond surety or other financial guarantee* "prior to the commencement of any mineral activities conducted pursuant to the general mining laws" which causes more than minimal disturbance to the environment. This means that if you want to work on a claim with anything more than a pick and shovel, you must post a bond. The bond amount must be "in an amount as determined by the Secretary of not less than $200 nor more than $2,500 per acre. ... Regardless of the financial limits of the preceding sentence, the bond, surety, or other financial guarantee shall not be less than the estimated cost to complete the reclamation of the disturbed land."

5. The Interior Appropriations Act, admittedly, still has to be approved by the House/Senate Conference Committee (may have been by the time this is printed), the House, the Senate, and the President. However, staffers (and several members of Congress) say this is about the best deal we can hope to get and that we may, in fact, get much worse. Remember, many of mining's radical opponents are members of the Conference Committee, and they will be attempting to load down the bill with every nasty thing they can dream up. The *House* version of the appropriations bill currently reads the same as the OMB's budget proposal on the $100 mining claim holding fee, so it is extremely likely that the Senate appropriations language will be accepted—if not something much worse.

Remember the Bureau of Land Management's proposal for rule-making on the posting of bonds or financial guarantees in August of 1991—for all mining activities that create land disturbances, including notice-level operations under the 5-acre rule? Remember the BLM's subsequent proposal in November 1991 to change the 5-acre rule, redefine "unnecessary or undue degradation of the environment," provide for additional criminal and civil penalties under the 43 CFR 3809 regulations, and other proposed revisions thereto?

Now, in the Department of the Interior Appropriations Act for FY1993, we have the entire packaged regulatory provisions firmly established in federal law, so there is no longer any need for the BLM to formulate a new set of 3809 regulations. For all practical purposes, the 5-acre rule has been rendered meaningless, all notice-level operations would require the posting of a bond or other financial guarantee and, in effect, unnecessary or undue degradation has been redefined as "more than minimal disturbance to the environment." So, through the budget appropriations process almost all of the BLM's most stringent rulemaking proposals would become fully realized, and the implementation of the Act's provisions requires the Secretary of the Interior to "promulgate rules and regulations" to carry out other provisions of the Act (that is, to fully implement the $100 mining claim holding fee and amendments to the mining claim patent process).

It is also interesting to note that there is no time frame specified for the financial guarantee provision. It seems likely, therefore, that bonding or financial guarantees will be required for existing notice-level or plan-of-operation-level activities at the time the FY1993 Interior Appropriations Act is enacted. In addition, consider this provision under the financial guarantee portion of the bill: "(d) Release.—The Secretary shall provide for public notice prior to any reduction in, or final release of, a bond or other financial guarantee." This provides an excellent opportunity for the filing of appeals or suits by environmental groups and may delay the release of bonds or surety indefinitely.

Let's assume the Act is approved "as is" (with nothing worse thrown into the package) and briefly summarize its impact on small-scale miners.

1. You pay the government $100 per claim next August, regardless of whether or not you did your assessment work before the Act was approved.

2. You pay the government an additional $100 per claim at the same time to hold the claim for the next year.

3. If you do any significant work on any claim at all, you will have to post a bond or other surety prior to the time the work is performed.

4. If the state in which you reside requires annual assessment work to be performed, you will have to post a bond or surety to

perform the work.

5. If you did assessment work prior to this September under a Notice or Plan, you will probably have to post a reclamation bond until the land disturbance has been fully reclaimed under the new requirements.

6. You still have to fully comply with all other laws, rules and regulations at the federal, state and local levels.

7. If you keep your claims, the number of on-site inspections will go up and compliance requirements will become ever more stringent.

8. If you pay the fees/post bond or surety/etc., you get absolutely nothing for it.

9. It will cost $110 to file a new claim ($10 filing fee plus $100 fee), plus the attendant state filing fees and taxes.

10. Under the conditions imposed by this legislative insanity, it is now estimated that well over 70% of the existing mining claims in the U.S. will be relinquished; grassroots exploration activity will almost cease to exist; and U.S. minerals production will decline dramatically within the next few years as deposits are mined out and no new discoveries are made to replace them.

22
Small-scale Miners Get Hit with $100 Fee and Bonding

IT IS NOW A CERTAINTY that miners will be required to pay a $100 mining claim "holding fee" per claim per year by August 31, 1993. Not only that, but they will be required to pay the next year's fees (from September 1, 1993, to August 31, 1994) in advance. So, if you're planning to hold on to your mining claims, you will have to pay $200 per claim next August 31. There is now no allowance for substituting assessment work in lieu of the holding fees, except for individuals holding 10 or fewer claims with less than 10 acres of total land disturbance. *However, miners must be operating under a valid notice or plan of operations and meet other specific criteria in order to qualify for the exemption.* If they meet these requirements, very small operations will be allowed the option of either paying the fees or performing and filing annual assessment work.

On October 1, Congress sent the final approved version of the FY1993 Interior Appropriations Act to President Bush for his signature, and there is no doubt that he will sign the measure into law. However, the provisions relating to the holding fee have been altered considerably from those that were originally proposed in the administration's budget. The language in the bill was amended, reamended and tinkered with numerous times by the House, the Senate and, finally, the House/Senate Conference Committee before the final compromise was approved and sent to the president.

For instance, the provision (called the "Stevens Amendment") defining the so-called small-scale miner's exemption now reads: "Provided further, That for fiscal year 1993, each claimant—(i) that is producing *under a valid notice or plan of operations* not less than $1,500 and not more than $800,000 in gross revenues per year as certified by the claimant from 10 or fewer claims; or—(ii) that is performing exploration work to disclose, expose, or otherwise make known possible valuable mineralization on 10 or fewer claims *under a valid notice or plan of operation;* and that has less than 10 acres of unreclaimed surface disturbance from such mining activity or such exploration work, may elect to either pay the claim rental fee for such year or in lieu thereof do assessment work required by the Mining Law of 1872 (30 U.S.C. 28-28e) and meet the filing requirements of FLPMA (43 U.S.C. 1744[a] and [c]) on such 10 or fewer claims and certify the performance of such assessment work to the Secretary by August 31, 1993."

The provision dealing with FY1994 reads exactly the same as the one noted above, except for the year it covers. It is especially significant that it also specifies "certify the performance of such assessment work to the Secretary by Aug. 31, 1993." This means that for this current assessment year (September 1, 1992, through August 31, 1993), *each claimant who wants to file the work instead of paying the fee must perform two (2) years' assessment work prior to August 31, 1993!*

Note that the work option is only available to those claimants operating under a "valid notice or plan of operation." So, any work performed (geochemical, geophysical, geomagnetic, surface sampling, etc.) that does not require at least a notice *does not qualify for the so-called small-scale miner exemption.* The requirement for a notice or plan is now especially significant because of the BLM rulemaking process.

The Bureau of Land Management (BLM) has announced that the agency will be issuing their final bonding regulations by late October or early November of this year. BLM officials say the new regulations will require bonding, surety or financial guarantees for all mining activities greater than "casual use." This means that, at an absolute minimum, some type of financial guarantee will be required for all Notice-level mining activities, and some small mining operations will be required to post a bond. *All* larger mining operations and exploration projects will be required to post a bond. So between the Appropriations Act and the BLM, where do we stand?

In order for a claimant to qualify for the small-scale miner exemption to the mining claim holding fee, an approved notice or plan of operations must be obtained, and a bond, surety or financial guarantee posted. Under current policy, a notice or plan is normally held open for from 3 to 5 years while the BLM conducts inspections on previously disturbed lands *after they have been reclaimed and reseeded.* In other words, the agency normally will not release a bond until after vegetation has been reestablished and sufficient time has elapsed to make sure there is no subsidence or problems with erosion in the reclaimed area. This means the ongoing costs of a bond, or the attendant financial liability associated with a financial guarantee, will go on for several years after the mined land reclamation has been completed.

It is also interesting to note that the holding fee provision is now set to expire at the end of FY1994 (September 30, 1994). In other words, the Appropriations Act only covers the next two fiscal years—there are no provisions for future years. This means that Congress has intentionally left this issue open for additional future legislative action, and there is no way that miners can predict what Congress may do with this matter in the next two years.

This introduces another critical element of uncertainty, and it makes future planning a practical impossibility. For instance, you may be stuck with an ongoing reclamation bond responsibility (meaning you have to hold the claims until it is released) while Congress decides to increase the holding fees or just extend them. Therefore, you may have to pay mining claim holding fees on lands that you have already reclaimed and in which you no longer have an interest in exploring or developing.

The provision for recordation of new mining claims states: "Provided further, that for every unpatented mining claim, mill or funnel site located after the date of enactment of this Act, through September 30, 1994, the locator shall pay $100 to the Secretary of the Interior or his designee at the time the location notice is recorded with the Bureau of Land Management to hold such claim for the year in which the location was made."

This means that you will pay $110 to record each new claim with the BLM ($100 holding fee plus $10 recording fee), in addition to the filing fees required by the state and county. There is no allowance for prorating the fees, so if you locate a claim in April of the next year you will still have to pay another $100 on August 31 (the same would apply if it was located in August 1993). Considering the economics of the process, it seems that anyone who locates a new claim prior to September 1, 1993, would have to be fairly well-to-do financially, or somewhat crazy.

The House/Senate Conference Committee also eliminated several amendments to the Interior Appropriations Act, most particularly those dealing with the mining patent issue. The House requirement for a one-year moratorium on the issuance of mining patents was deleted, as was Sen. Reid's amendment to require payment of full market value of the land for mineral patents and other changes in mining patent law. This might seem a little strange because of the furor over the patent issue, but it has a logical explanation. You see, the public and the media have been fed so many lies about mining patents that the issue has become a primary tool used by radicals and anti-mining groups in their attempts to abolish the Mining Law of 1872, as amended. Therefore, mining's enemies weren't about to allow the approval of legislation that would defuse the issue, because they fully intend to use the patent issue to achieve their ultimate objectives. By deleting Sen. Reid's patent amendment, they can continue to feed propaganda to the general public and thereby gain more support for mining law "reform."

The Conference Committee also deleted the so-called Reid-Bumpers amendment, which required a bond, surety or financial guarantee for all mining activities "causing more than minimal disturbance to the environment." This change is moot, however, because the BLM intends to implement their new bonding regulations as soon as possible. Recent discussions with BLM officials revealed that the agency has applied to the Office of Management and Budget (OMB) for a waiver from the regulatory freeze imposed by President Bush and the Administration, thereby allowing the BLM to publish the new regulations in the Federal Register in the very near future.

Another change in the Act is the addition of another provision, which states: "Provided further, that for purposes of determining eligibility for the exemption from the claim rental fee required by this Act, any claims held by a husband and wife, either jointly or individually, or their children under the age of discretion, shall be counted together toward the ten claim limit."

This provision is obviously intended to close a potential loophole whereby a family could divide up a group of claims under the individual name of each family member and still meet the 10-claim limitation for the mining claim holding fee exemption. As mentioned in the first paragraph, the first mining claim holding fee provision requires the payment of $100 per claim on August 31, 1993, for the current assessment year (September 1, 1992, through August 31, 1993).

The second payment is required by another provision, which states, in part:

> ...each claimant shall, except as provided otherwise by this Act, pay an annual claim rental fee of $100 per claim to the Secretary of the Interior or his designee on or before August 31, 1993 in order for the claimant to hold such unpatented mining claim, mill or tunnel site for the following assessment year beginning at noon on September 1.

So, $200 per claim must be paid to the BLM next August 31, or the claims will be invalidated.

Recent discussions with Nevada Department of Minerals officials, who have been surveying large and small mining companies as well as small miners in regard to the imposition of a mining claim holding fee, indicates that the $100 fee provision will produce an overall reduction of at least 50% in the total number of active mining claims, with small-scale miners dropping up to 70% of their total claim holdings.

Most industry officials, mining professionals and small-scale miners agree that the $100 fee will severely impact minerals exploration in the U.S. It is also expected to eliminate a large number of small-scale miners, mining professionals, independent consultants and smaller exploration firms from participation in this country's mineral exploration and development, thereby producing a greater reduction in employment and economic activity than could possibly be offset by revenues to the U.S. Treasury produced by imposition of the fee.

In addition, the new BLM bonding, surety or financial guarantee requirements for Notice-level and small mining operations is expected to have a considerable impact on mineral exploration activities, especially for the smaller exploration firms and individual small-scale miners and prospectors.

It should also be kept in mind that the BLM is still processing and finalizing the proposed rules for revising the 43 CFR 3809 regulations to provide for possible modification or elimination of the 5-acre rule, as well as the other proposed changes discussed in field hearings last year. This rulemaking process is entirely separate from the bonding regulations, and it has the potential to produce additional negative impacts on U.S. minerals exploration and small-scale miners.

23
Locating and Recording New Mining Claims

CMJ Editor's Note: The California Mining Journal made every possible effort to obtain information concerning the new $100 mining claim holding fee regulations for publication in the August 1993 issue of the CMJ, in order to inform readers of the new requirements in a timely fashion—particularly in reference to the small miner exemption. However, the new regulations were not published in the Federal Register until Thursday, July 15, and were not distributed to the general public until the following week—well after the publication deadline for the August issue.... Therefore, this item will only address the filing of new mining claims and the new requirements as of September 1, 1993.

ACCORDING TO THE NEW RULES and regulations found under Part 3830—Location of Mining Claims, section 3833, 1-5(a): "Mining claim or site (mill site or tunnel site) located on or after October 6, 1992, and on or before September 30, 1994. The $100 non-refundable rental fee, for the assessment year in which the mining claim or site was located, shall be paid for each mining claim, mill site or tunnel site at the time of recording of the mining claim, mill site or tunnel site pursuant to section 314(b) of FLPMA and section 3833.1-2 in addition to the service charge required by section 3833.1-4(a)."

This means that all new mining claims located during the time period specified above are subject to the $100 nonrefundable rental fee plus a $10 service charge at the time the Certificate or Notice of Location is recorded in the BLM State Office. For those claims located prior to August 31, 1993, and recorded after August 31, 1993, the fee will be $200 per claim plus the $10 recording fee. (Most claim holders pay fees for both years.)

Under these regulations, when new mining claims are located and recorded the full rental fees (no prorating) must be paid and there is no provision for a small miner exemption on new claims (even if you are currently operating under an approved exemption). So, all new claim recordings with the BLM must be accompanied by a payment of $110 at the time of recording, and, if the date of location and time of recording span two assessment years, then a payment of $210 per claim is required. Keep in mind that the new federal requirements do not supersede state law—so claimholders must meet all state and county filing requirements and fees in addition to meeting the requirements of the new BLM regulations.

Because of conflicts created between state and federal requirements by the $100 rental fee legislation and regulation, many states have enacted new statutory requirements for mining claim staking, filing and recording (including, in some cases, increases in recording fees as well as new procedures for recording at the county level) in order to comply with federal law while still meeting individual state statutory compliance. Therefore, all miners should recheck the state requirements for locating and recording claims in the state in which the claims are situated (there are numerous variations from state to state which pertain to this matter).

The law that created the $100 rental fee requirement (the current law) expires on September 30, 1994; consequently, there is another important factor to keep in mind: If a mining claim is located prior to August 31, 1994 and recorded with the BLM after September 1, 1994, a fee of $210 will be charged at the time of recording ($100 for each of two assessment years). If claims are located during the period between September 1, 1994, and September 30, 1994, a fee of $110 will be required at the time of recording with the BLM—even if the claim is recorded after October 1, 1994.

Also note that the new federal requirements described above are, in all probability, temporary until Congress enacts some type of mining law reform. In addition, legislation to make the $100 fee permanent is currently being considered. So, it is quite likely that the new federal statutory and regulatory requirements may be changed before the end of this year, and it is almost a certainty that they will be changed prior to the expiration of the existing enabling legislation.

In other words, the foregoing information deals exclusively with the requirements of the new regulatory regime as they relate to implementation of the $100 mining claim rental fee, and these requirements are likely to change within a fairly short period of time. The location and recordation of mining claims under federal regulations may also undergo extensive revision as a result of potential changes in the BLM regulations, a number of which have been under consideration for quite some time (e.g., proposed rulemaking to revise the BLM's 43 CFR 3809 regulations and the potential issuance of the bonding and financial guarantee regulations).

Because everything related to mining in the United States is still in a state of movement and change, miners and the mining industry cannot look forward to any stability or dependability in the federal laws, policies and regulations for even fairly short periods of time. Therefore, miners should be aware that even new mining laws and regulations may become obsolete practically overnight.

Special Note: In addition to the possible changes to the $100 holding fee requirement mentioned above, the Omnibus Budget Reconciliation Act of 1993, which received approval on August 6, contains significant revisions in reference to the $100 mining claim holding fee as well as the location fee charged by the BLM at the time a Certificate or Notice of Location is recorded. The new "location fee" required is set at $25 per claim (replaces the $10 fee per claim noted above), and the $100 maintenance fee is extended through the years 1994–1998. In the event that these fee requirements are not included in an overall "mining law reform" package, it is almost certain that the above-mentioned changes will be incorporated into the budget appropriations process.

24
New Mining Claim Fee Policy Implemented

Author's Note: It is important to point out the fact that the recent changes in federal laws, rules, regulations and policies related to mining claims *do not affect state laws and regulations in any way.* In other words, the state mining statutes and regulations must still be complied with by all mining claimants. Because of the U.S. Government's implementation of the new claim fees and requirements, many of the individual states have also changed their statutes and regulations this year in order to make them compatible with the changes in federal laws. All mining claimants must therefore check their respective state requirements to ensure compliance.

JUST WHEN YOU THINK you've finally understood the requirements of a new law or regulation, the government either changes it or enacts a new law and/or issues a new set of regulations. The regulations that were published in the July 15, 1993, Federal Register concerning implementation of the mining claim holding fee were modified, added to, and extended just 26 days after they went into effect!

When President Clinton signed "The Omnibus Budget Reconciliation Act of 1993" (H.R. 2264) on August 10, 1993, it took effect immediately. Thus, the $100 mining claim holding fee (now called "Claim Maintenance Fee") provision was extended through 1998: a new $25 "location fee" per claim was added (in addition to the $10 recording fee); the 10-claim small miner exemption (now called "waiver") was modified; the Secretary of the Interior was directed to adjust (read: increase) the required fees to reflect changes in the Consumer Price Index every 5 years or more frequently; and the Secretary was directed to develop a new set of rules and regulations.

Because of the difference between the date of enactment (August 10, 1993) and the end of the assessment year (now August 31), there are now two possible mining claim fee schedules at this time:

1. A claimant who locates a claim between August 10, 1993, and August 31, 1993, and who records the claim after August 31, 1993, must at the time of recording pay the $10 service fee, the $25 location fee, one $100 fee (to hold the claim for the balance of the assessment year through September 1, 1993) (total: $135 to BLM), and unless qualified for the small miner or other exemption, a second $100 fee (to hold the claim for the assessment year of September 1, 1993,

through September 1, 1994) (total: $235 to BLM). The claimant would also pay $100 on or before August 31, 1994, to hold the claim for the assessment year of September 1, 1994, through September 1, 1995, and $100 each August 31 from 1995 through 1998, plus all state fees.

2. A claimant who locates and records a claim between September 1, 1993, and August 31, 1994, must at the time of recording pay the $10 service fee, the $25 location fee, and a $100 maintenance fee to hold the claim for the balance of the assessment year through September 1, 1994. The claimant would also pay $100 on or before August 31, 1994, to hold the claim for the assessment year of September 1, 1994, through September 1, 1995, and $100 each August 31 from 1995 through 1998, plus all state fees. Total: $135 per claim to BLM at time of recording, unless the location date is in the assessment year prior to the assessment year in which the claim is recorded; then the total fees would be $235 as above.

At this time, the Act (P.L. 103-66) expires on September 30, 1998. There are no provisions for prorating the fees, so if you locate a claim one day before the end of the assessment year you must pay $200 in claim maintenance fees for both years. Once paid, no fees will be refunded, even if you immediately quitclaim the claim or claims.

It is also important to understand that the current efforts to enact mining law "reform" in the U.S. Congress may add more fees to those noted above. You see, the current $100 fee requirement is in lieu of federal annual assessment work requirements, but several of the mining law bills in Congress also propose a mining claim "rental fee" in addition to other types of "administrative fees."

PART IV

Nevada Mining's Fight to Survive and Thrive

"...the hard-fought battles for fair and reasonable treatment of miners and the anger aroused by the frequent demeaning and derogatory statements made by certain politicians and legislators will not be easily forgotten. Miners have become fed up with the constant trashing of mining by public officials and the news media.... After all, how often do miners and the industry as a whole get publicly trashed in a State of-the-State address by a new governor?"

25
Nevada's Gold Mining Industry

DESPITE THE DRAMATIC decline in most U.S. metal prices and production over the past few years, the value of Nevada's mineral output continued to increase. This is primarily due to the tremendous expansion taking place in the state's gold mining business, which also accounts for the fact that the state is now producing well over 50 percent of the nation's total annual gold production from domestic mines.

Nevada has been a leader in the new gold rush of the 1980s, and the state is currently attracting worldwide attention as a prime gold exploration target. Although a far cry from the California Gold Rush of 1849, the steady influx of foreign and domestic mineral exploration and development companies has created a multi-million-dollar mining exploration business in the state within just a few years.

The old-time mining booms in Nevada, such as that created by the discovery of the Comstock Lode at Virginia City, resulted mainly from the discoveries of fairly rich gold and silver ores occurring in veins, most of which outcropped at, or close to, the ground surface. In contrast, most of the state's current large gold mining operations are mainly based upon huge ore deposits that contain extremely fine, disseminated gold particles. Most of these ore deposits were not exposed at the ground surface, and they were only discovered and defined through the use of sophisticated exploration technology and modern drilling techniques.

The micron gold particles in the ores are often only visible under a microscope, so their presence was not even suspected by the earlier miners. The presence of gold in these types of ores can usually be determined only by assay, since most of the minable ores are nearly indistinguishable from the adjoining waste rock.

The unique character of most disseminated gold ores has brought about a revolution in prospecting techniques, exploration methodology, mining procedures and processing technology. Nevada's gold mining industry has become a world leader in the use of innovative, cost-effective and efficient applications of the most advanced mining and processing technology available. One example of this is the fairly recent development of cyanide heap-leaching technology for processing large quantities of low-grade precious metal ores. This modern, low-cost processing technique makes possible the profitable recovery of gold and silver from ores previously thought to be worthless, and it has contributed significantly to the large increases in the state's gold production.

However, even with the tremendous technological advances that have been made in the industry, modern gold mining can still be a somewhat risky and expensive business. It can be very difficult for mine managers to prepare adequate long-range business plans when the gold market is highly volatile, as it has been several times during the past few years. This condition creates a rather unique business environment, wherein the higher-grade ores (most profitable) are usually mined when gold prices are lower and the lower-grade ores (least profitable) are mined when prices reach a higher level.

This type of operational procedure can decrease the potential for higher short-term profits, but it increases the overall life of the mining operation and, as a result, produces a much higher profit over the long term. It also encourages hedging in the gold futures market, with a certain amount of production sold forward at a set price to smooth out any impacts resulting from short-term fluctuations in the gold price.

Many large mining companies have established exploration firms and offices in Nevada, as have a multitude of smaller companies, and their numbers are still increasing. Although a few of these firms do not as yet have operating mines in the state, their exploration and development activities generate a significant inflow of investment capital—which, in turn, boosts the state's economy and provides additional local employment.

In addition, intensive mineral exploration is absolutely essential to a healthy mining industry. The ongoing discovery of new mineral resources and the definition of additional ore reserves are required to sustain the existing production capacity and to provide for expansion in the industry. Mineral exploration today, however, is a very expensive business.

A mining industry official recently said that it costs around $250 million to discover and develop a world-class gold deposit, adding that it is not currently worthwhile to develop most base metal ore deposits, unless they contain a fairly high grade of minerals, because of the current low prices paid for most metal commodities.

The state of Nevada's revenues from the gold mining industry have increased steadily over the past nine years and received a considerable boost from the recent tax legislation enacted by the 1987 Legislature. In addition, the industry also contributes a significant amount of capital to the local economies throughout the state, since most of the major producing mines are located in rural county areas. Many of Nevada's smaller towns are dependent upon mining as their primary source of revenue, employment and economic activity.

The state's mining industry is now second only to gaming and tourism in terms of size, gross revenues and importance to the economy. The total value of mineral production is continuing to increase at a rapid rate, and the mining sector has emerged as Nevada's largest basic industry.

Contrary to some commonly held misconceptions about mining, it is a business much like any other basic industry. Unrestricted or irresponsible mineral development activities are a thing of the past, and today's modern mining industry utilizes the most efficient technology available to guard against any undue degradation of the natural environment. Surface reclamation plans are a necessary and integral part of operational planning, and all of the state's operating mines have ongoing programs for reclamation and restoration of disturbed lands as the mining operations progress.

The number of new gold mines that have reached the production stage has been steadily increasing every year since the early 1980s. In addition, several new gold mining projects, as well as expansion programs for some of the operating mines, are currently being planned in the state. The current gold mining boom in Nevada is in full swing, and it is expected to continue for many more years.

Overall, Nevada's gold mining industry is poised for rapid growth and a bright future.

26
Nevada's Proposed Gold 'Fee' Opposed

THE NEVADA STATE LEGISLATURE became the focus of considerable concern to the mining industry and rural residents of the state when, on February 6, 1987, Assemblyman Marvin Sedway introduced his "Jim Dandy" bill to assess a "fee" of $16.50 per troy ounce of gold (or a portion thereof) extracted from mines in the state of Nevada.

In his introduction of Assembly Bill 161 (A.B. 161), Sedway said his proposal is simply "a fee to do mining in this state" and maintained that the assessment is not a tax on the mining industry—despite the fact that it specifically targets gold mining and the monies generated would be used as general revenues for the state. Just prior to his introduction of the bill, Sedway had stated his intention of raising welfare payments and giving a raise to teachers in the state from revenues obtained by additional taxation of the gold mining industry.

While attempting to justify his action, Sedway claimed that mining companies in Nevada were based in locations outside the state and are "paying nothing while they rape the state of Nevada." The Las Vegas Democrat obtained 26 co-sponsors of the bill, most of whom are from the southern part of Nevada where there is not much gold mining.

Sedway said that he decided to change his "Jim Dandy" mining severance tax into a mining "fee" because it was the "method of least resistance." The state already charges a mining severance tax on net proceeds of mines for precious metals extracted in Nevada and, Sedway said, if he had tried to place another tax on the gross revenues of the mining industry it would have been declared unconstitutional.

Opponents of A.B. 161, however, maintain that the fee is actually a tax on gross production regardless of what it is called. In addition, the bill singles out one segment of the mining industry alone, largely because gold mining is the only metals-producing sector of the industry that is still surviving and making a profit in the state. Most of the other metals-producing mines in the region have been forced to close because of depressed prices and intense foreign competition.

According to a recent legal opinion expressed by a Washington, D.C.-based law firm, Convington & Burling, Sedway's A.B. 161 claims to impose a per-ounce fee on gold but it is "clearly a tax designed for general revenue purposes." As a result, the law firm says the bill would produce double taxation on mines and is unconstitutional. The former chief legislative lawyer for the state of Nevada, Frank Daykin, has also expressed an opinion that the gold fee is unconstitutional.

The initial debate over the constitutionality of the measure began when several state assemblymen objected to the bill being sent to the Assembly Natural Resources, Agriculture and Mining Committee, rather than the Assembly Taxation Committee. In a move to bypass the normal bill-processing procedures in the Assembly, Sedway obtained enough support to force an immediate vote to send the bill to the Natural Resources Committee, where six of the eleven members are co-signers of his bill.

The committee selection process "abrogates the system of the legislature," said Assembly Minority Leader Lou Bergevin, R-Gardnerville. "A fee is a tax. It's going to the wrong committee."

"I have real problems in singling out an industry for a fee that is really a gross tax," said Senator Ken Redelsperger, who chairs the Senate Taxation Committee. He said that Sedway is "really trying to circumvent the (state) constitution" by calling for a fee and not a tax.

"Sure it's a tax," Redelsperger said. "He may be calling it a fee, but you usually get a service for a fee." He has ensured that the Senate Taxation Committee will hold hearings on the bill if it is passed in the assembly, even though he already believes the bill is unconstitutional.

Just prior to the first hearing held on March 16, the mining industry invited the Assembly Natural Resources, Agriculture and Mining Committee to a tour of several Nevada gold mines in order to inform the legislators about mining operations and the industry's economic contribution to the state—especially in rural Nevada. The committee toured Freeport Gold Company's Jerritt Canyon Mine near Elko and the Echo Bay-Homestake Round Mountain Mine in central Nye County, and met with local government officials, community leaders, and representatives of the mining companies.

The hearing on March 16 opened to a packed and emotionally charged audience, with Assemblyman Sedway leading off with his testimony for the first hour. He immediately attacked the mining industry, saying miners were not paying their fair share of taxes, that the industry was threatening their employees and rural communities, that mining was raping the state of Nevada, and that there would be no effect on gold mining if his bill is passed. He also said the mining companies were all based in out-of-state locations and that they were lying when they said that mines would be closed and the industry would move out of the state.

His main point, however, was that gold mining had to stay in Nevada because that was where the gold was. "Where are they going to go?" Sedway said, "Iowa? Georgia? New York? Maine?" He said the companies were "terrorizing" rural communities and employees with threats of closures and layoffs if the bill passed.

In the subsequent testimony opposing A.B. 161, any individual appearing before the committee who happened to challenge Sedway or any of the bill's supporters was immediately chastised by Assemblyman James W. Schofield, the committee chairman, and several other committee members—all of whom are co-sponsors of the bill. Critical and demeaning testimony against the gold mining industry was allowed expression by all who appeared in support of the bill, while any response to the allegations, or individuals making the allegations, by opponents of the measure was immediately challenged and disallowed.

In spite of this obvious bias against mining, a large number of industry representatives and individual miners continued to present factual testimony that showed how the bill would adversely impact the state's gold mining industry. In addition, mining officials presented detailed evidence that gold producers were, in fact, already paying substantial taxes in Nevada, through the existing sales, property and net proceeds of mine taxes. Each presentation by miners, though, was eventually challenged by committee members in an attempt to show that gold producers were capable of paying higher taxes.

The mining industry repeatedly pointed out that only mining and gaming industries in Nevada currently pay any form of tax on income, while other businesses and the state's residents are not subjected to taxes on either net or gross income. As a result, they said any additional taxes on these two businesses would be an unfair and unequal burden. Unfortunately, the present inequalities in the state's system of taxation did not appear to interest the members of the committee, and they persisted in their efforts to show that gold producers were not paying their "fair share" of taxes. Nobody, however, seemed able to define exactly what the term *fair share* meant, and the committee refused to consider that the same formula could be applied to other businesses in the state.

The primary supporters of the gold tax that appeared before the committee were the State of Nevada Employees Association, the Nevada State Education Association and welfare proponents—all of whom said they needed a raise in salaries and welfare benefits. They were unanimous in their opinion that the gold mining industry was the best available source of revenues to provide the funds necessary to give them a raise.

The opponents of A.B. 161 represented a broad cross-section of the state's rural residents, chambers of commerce, county commissioners, city officials, local businesses, state senators and assemblymen, attorneys, mining industry officials, and the state's small miners and prospectors. All of the individuals and company representatives presented testimony that showed the bill would have a negative impact on gold mining in Nevada and stressed the following points.

If passed by the legislature, the gold tax would:
- Reduce the profitability of gold mines.
- Reduce the net proceeds of mines revenues to the state and counties.
- Discourage or eliminate many new mining operations.
- Severely reduce exploration for minerals in the state.
- Cut future mine employment and cause layoffs of current mine workers.
- Force many marginal mines to close.
- Reduce investment capital available for mineral development.
- Cause many mining firms to close exploration offices and move out of state.
- Reduce operational life of gold mines by lowering the profitability of mining and processing lower grade ores.
- Reduce economic activity and tax revenues in rural counties.
- Reduce gold production and increase reliance on foreign imports.
- Effectively eliminate small miners and prospectors in the state.
- Reduce the number of mining claims and filing fees paid.

- Adversely impact businesses that service and supply the mining industry.
- Reduce the state's ability to attract new business and industry.
- Hamper efforts to diversify the state economy.

In addition, opponents pointed out that the gold fee is a tax designed to produce revenues to support state programs that are not caused by or related to mining activity; that it adds a tax on gross production to the existing net proceeds of mines tax; that it singles out one sector of industry for taxes without taxing other businesses or residents equally, and that it disregards the profitability of each business operation subjected to the tax.

As an alternative to the gold tax, several opponents of the measure suggested the legislature raise the current sales or property tax rates. They felt that if Nevada needs more tax revenues, then the burden should be shared equally among all of the state's businesses and residents who will derive benefits from the expenditure of those revenues.

Because of the large number of people who wished to present testimony in opposition to A.B. 161, the hearing on March 16 was held over for an additional evening session (until about 10 p.m.). By that time, many of the committee members had left the hearing and those in attendance were tiring. As a result, the hearing was extended to a future date, at which time further testimony will be presented.

This hearing before the Assembly Natural Resources, Agriculture and Mining Committee is expected to be just the first in a series of hearings on Nevada's proposed gold tax. If the measure is passed out of the committee and passed by the state assembly (which seems likely under the prevailing political climate), then A.B. 161 will still face hearings by the Senate Taxation Committee before it comes to a vote before the senate.

Many of the individuals close to the issue believe the bill will be defeated in the senate, although they say the measure could still become law given the current anti-mining attitude being expressed in the state. Further details concerning the impending hearings and the final outcome of the issue will be reported in the near future.

27
Nevada's Mine Tax Issue Still Undecided

AS THE 1987 NEVADA STATE legislative session approaches its final days, the issues pertaining to increases in the state's mine taxes still remain unresolved. Most informed sources expect a political battle over the type and amount of tax imposed on the industry as the session comes to a close. Regardless of the immediate outcome of the hotly contested debate, mining officials predict that the issue will remain a political football over the next few years.

Nevada's miners have agreed to support the enactment of Senate Joint Resolution 22 (S.J.R. 22), which provides for an increase in the taxes paid on minerals extracted in the state to roughly two-and-a-half times the present level. The new tax revenues generated by the resolution can start flowing into the state's coffers as early as July 1989. (Because the proposal would amend the Nevada Constitution it must be enacted by the 1987 and 1989 legislatures and also be ratified by the state's voters during special elections held in conjunction with municipal elections in the spring of 1989.) The mining industry has also offered to aid in passing the mining tax increase by committing the necessary resources to educate the general public about the benefits of S.J.R. 22.

In the interim, Joe Murray, president of the Nevada Mining Association and president of Freeport Gold Co., has informed the Senate Taxation Committee that the association has drafted legislation to provide about $20 million from the industry to be utilized by the state during the period before the resolution takes effect. The additional money offered is designed to help state government meet a projected shortfall in its budget revenues over the next two years.

The amount of the payment was agreed to during several meetings between legislative leaders and industry officials. However, an assembly taxation subcommittee composed of three Las Vegas Democrats openly rejected the mining industry's offer and proposed instead that the state's gold mining industry give the state a $30 million "grant" over the next two years. Assemblywoman Myrna Williams' subcommittee felt that miners should "give" the state $15 million each year (in addition to the current net proceeds of mines tax) until the plan to increase net proceeds of mines taxes is passed. One legislator has termed the subcommittee's proposed plan "blackmail."

The proposed $30 million gift to the state over the next two years would represent a portion of the mining industry's future tax payments if S.J.R. 22 is passed. However, Assemblyman Bob Gaston, D-Las Vegas, said the advance tax payments "would be a loan that does not have to be repaid because the trigger mechanism for repayment is so high that it never kicks in." The subcommittee's proposal is $10 million higher than the previously agreed two-year advance tax payment plan that is also tied to S.J.R. 22.

Williams, D-Las Vegas, said the miners' offer of $20 million "is not sufficient, considering certain overruns in our budget." She said the subcommittee "put forth a real effort to be fair" by not pushing ahead with another proposed $42.5 million annual mining tax plan.

Assemblywoman Gaylyn Spriggs immediately objected to the subcommittee's tactics. Spriggs, R-Hawthorne, said, "The industry was told they were going to get a petition for a severance tax if they didn't cough up some money. Now, that's blackmail." The Las Vegas Democrats threatened to ask the state's voters to approve a huge severance tax measure on mining and minerals if their legislation isn't passed during this session.

Spriggs added that "it's not right for the state to say it has a shortfall, and (then) have one industry pick up the whole shortfall."

The existing Nevada tax on net proceeds of mines is limited by the state constitution to the various property tax rates levied by the counties. Last year, the statewide average county rate of 1.63 percent generated $6.1 million in net proceeds of mines taxes for the counties. The mining industry had proposed to increase the rate of taxation to 5 percent several years ago, but that measure was defeated by the state's voters. S.J.R. 22 would amend the constitution to boost the state's net proceeds of mines tax to $5 per $100 of assessed valuation (or 5 percent), which is much the same as the mining industry's original proposal for increasing mine taxes. The resolution also guarantees that the counties will continue to receive their share of the tax, based upon their current property tax rates. The additional tax revenues would go into the state treasury and, based upon the current property tax rates, is expected to generate about $21 million each year for the state's general budgetary requirements.

Joe Murray said, "If mining continues to expand in Nevada, the constitutional amendment will generate substantial new tax revenue for the state."

Murray also explained the industry's offer to provide $20 million in extra revenue prior to the enactment of S.J.R. 22. The healthiest segment of the mining industry would make an extra $10 million prepayment of net proceeds of mines taxes during 1987; normally these taxes would be paid in July of 1988. All of this money would be available for the state's 1987-88 budget. The most profitable mining companies have also agreed to advance another $10 million prepayment of taxes for the state's 1988-89 budget, but the 1988 payment would be based on future anticipated tax liabilities.

He said, "This voluntary effort provides important funding for the state's use during the current budget shortfall, until long-term revenue needs are resolved." He added that industry officials are looking forward to participating and assisting in the tax study currently being discussed in the legislature, which is designed to define Nevada's future budgetary requirements.

Murray stated that a bipartisan committee vote for S.J.R. 22 would benefit the long-term viability of the mining industry and the citizens of Nevada. He said, "The mining industry recognizes the state needs additional money. We are good corporate citizens, and want to continue doing business in Nevada."

Murray also said the resolution "will increase mining's net proceeds taxes an average of two-and-a-half times its present level for all profitable mines in Nevada." As an example, he noted that S.J.R. 22 could raise about $28 million in annual taxes for the fiscal year 1988-90. "No other industry in Nevada's history has stepped forward with such a proposal for a major tax increase," he concluded.

However, Assembly Taxation Chairman Paul May, D-North Las Vegas, said that while the state's mining industry has made a generous offer, "In my opinion, I think they could go further." He said he's delighted that the subcommittee came to some sort of agreement on how to raise more money.

Despite the fact that Nevada's mines produce a wide variety of mineral materials, the assembly is targeting the gold mining industry for the $30 million "gift" to the state because it is the only metals-mining sector of the industry that is currently making a substantial profit. Most of the miners feel that this singling-out of one mining sector to provide increased revenues for the state is unfair and unjust—if not unconstitutional.

Because of this, and the attempt to raise the ante by another $10 million, informed sources expect a legislative dogfight over the mining tax issue during the last hours of the current legislative session. Even if some type of compromise is reached between the Nevada assembly and senate, it looks like the state's gold mining industry has been targeted for future attention by the Nevada State Legislature.

28
Nevada Mining Tax Bill Signed into Law

NEVADA GOVERNOR RICHARD BRYAN signed a bill on June 26, 1987, allowing the state to accept $20.5 million in advance tax payments from the mining industry, based upon the expectation that a constitutional amendment authorizing the levy will be approved by both the 1989 Legislature and the voters of Nevada. The signing of the legislation into law ended a controversial, sometimes heated battle in the Nevada State Legislature over just how much tax revenue should be required from the state's mining industry, the rate of taxation on the proceeds of mines, and how those tax revenues should be divided between state and county budgetary funds.

Under A.B. 872, the mining industry will make a $10 million "prepayment" of taxes in July 1987 and another $10.5 million for 1989, based upon the anticipated revenues from a revised net proceeds of mines tax. The compromise legislation is predicated on legislative and voter approval of a proposed constitutional amendment outlined in S.J.R. 22, which will allow the state to tax mines at a rate of $5 per $100 of assessed valuation. S.J.R. 22 requires legislative approval this year and also in 1989 before it can be presented to the state's voters.

Under the measure, when the state has collected $50 million from the net proceeds of mines tax it is then obligated to credit the full $20.5 million against future tax payments by the industry. This is not expected to occur until about 1992.

Nevada Mining Association President Joe Murray said in a recent news release, "Nevada's mining industry came forward with a substantial contribution to the state in its time of need. Some of the alternatives which were discussed could have seriously hurt the mining industry in the long term."

The existing net proceeds of mines tax averaged about $1.80 per $100 of assessed valuation. The tax is based on the net profits of the individual mining companies, and the revenues from the tax previously went only to the county in which the ore was mined. Since the state of Nevada needed additional tax revenues to balance its budget over the next two years, an advance payment program was devised by mining industry officials and some of the leading legislators to allow for the collection of tax prepayments pending approval of the necessary constitutional amendment. The battle lines were drawn, however, when a bloc of Las Vegas legislators introduced a series of bills designed to obtain the maximum amount of revenue possible from the healthy segment of the gold mining industry, arguing that the gold miners were not paying their "fair share" of taxes to the state.

Among the most prominent of the initiatives that Joe Murray referred to was Assemblyman Marvin Sedway's "Jim Dandy" gold fee, which was bitterly opposed by both rural legislators and the state's miners. The proposed bill, A.B. 161, would have imposed a "fee" of $16.50 (later amended to $11) per troy ounce on every ounce of gold extracted in Nevada as a payment (according to Sedway) "for the privilege of doing business in Nevada." Despite some very demeaning remarks about the mining industry in general and some specific threats to both the Nevada State Senate and miners, Sedway's bill was eventually killed by the Senate Taxation Committee.

This proposal was immediately followed by several companion measures drafted in the state assembly, one of which would have removed the concept of taxation on the net proceeds of mines and substituted instead a method for taxing the gross proceeds of mining operations—disregarding whether or not any particular mining operation was, in fact, producing a profit. Another measure attempted to eliminate many of the currently legitimate deductions allowed in the computation of net proceeds, including disallowing any deductions for other taxes paid to the state. According to mining industry officials, many of these measures, if passed, would have effectively crippled many mining operations in the state and caused a number of mining companies to seek greener pastures elsewhere.

The original version of A.B. 872 would have, in effect, required the state's mining industry to provide a "gift" of $30 million to the state, both by increasing the amount of the tax "prepayments" and raising the necessary amount of tax received by the state to such a high level that the advance payments would never be credited towards future tax liabilities. The $20.5 million in advance mining taxes was approved after a compromise was reached between mining industry officials and several of the assembly leaders in a behind-the-scenes meeting. A downward revision in the amount of tax revenues received by the state from net proceeds of mines before the mining industry could receive credit for the tax prepayments was also worked out during the closed-door meeting.

Following the governor's signing of the mining tax measure, the state's mining industry published an advertisement in local newspapers that reads, in part:

> Under this law, Nevada's mining industry will make a $20.5 million prepayment of taxes to the State of Nevada during the next two years. The money will be used by the state to pay for education and other vital public services. The Nevada mining industry came forward with this tax prepayment offer to help bridge the gap before equitable, long-term state budget solutions could be reached.
>
> As for the mining industry's view of an appropriate long-term solution to the state's revenue dilemma, we are actively supporting an amendment to the Nevada Constitution calling for a permanent mining tax increase. This increase is at least 2.5 times the mining industry's present tax level. No other industry in Nevada's history has come forward with such a dramatic, major tax increase. The amendment passed the legislature this session. If it passes again in the next session of the legislature, Nevada voters directly will get a chance to decide on the plan on a ballot issue as early as 1989. We hope Nevadans will support the concept of the tax increase....
>
> Developing this mining tax package was no easy process. It required months of carefully designing a tax plan that considered the ups and downs of world mineral prices and many other concerns. Some of the alternatives discussed could have seriously hurt Nevada mining in the long term. ...
>
> For more than a century, mining has served as a backbone industry for the state's economy. In the last decade, the mining industry has invested more than a billion dollars to bring mines into production. The industry continues to be the economic stabilizer for most of the rural communities in Nevada. in Nevada. Today, mining provides thousands of jobs directly and in support services. ...
>
> We hope mining in Nevada can continue to grow and contribute toward building a stronger economy in our state. The mining industry extends a hearty thanks to the many friends, suppliers, employees, community and business leaders, patrons, and neighbors who have provided the opportunity for continuing a proud heritage. ...
>
> We will continue to do our part.

Nevada's new mining tax plan was a hard-fought (and sometimes bitter) issue that involved a considerable amount of intense, behind-the-scenes activity, even (at times) threatening to produce a dramatic north-south split in the Nevada State Legislature. But, as the saying goes, "All's well that ends well."

29
Anti-Mining Forces Focus on the Comstock Lode

AFTER 138 YEARS OF ALMOST continuous mining activity on the Comstock Lode, which includes the famous silver bonanzas mined at Virginia City, Nevada, the Comstock Era seems to be coming to an end. Paradoxically, depletion of the area's precious metals deposits is not the reason behind the closure of this historic mining region: anti-mining interests have targeted the area for an historic district, complete with a proposed view shed in which no unsightly mining activity is to be allowed.

The Comstock Historic District was established in an effort to preserve the district's scenic and structural qualities as reminders of Nevada's first major mining boom, and the colorful and hectic days that led to statehood during the Civil War. The region is literally covered with old mine workings, mills, dumps and tailings piles left behind by the early miners.

In recent years, the mining community's old-timers have been gradually replaced by a new breed of residents: outsiders who moved to the area for its scenic qualities, solitude and historic mining flavor left over from the boom days at Virginia City, Gold Hill and Silver City. Many of the local businesses now ply the tourist trade that is attracted by the region's historic qualities. As a result, most local residents are employed in tourist-related activities or they work at jobs in Reno or Carson City, commuting daily from their homes on the Comstock.

This non-mining population has now effectively taken control of the area, and they vehemently oppose any and all new mining activity as soon as it is proposed. The residents have also enlisted the aid of various preservationist, conservationist, historical and environmental groups in their efforts to shut down mining throughout the entire district. Their message is clear: old-time mining with its historical flavor is absolutely wonderful, while modern-day mining activity is dirty, dangerous and destructive.

The Comstock Lode district overlaps two Nevada counties: Storey County and Lyon County. Both counties have fairly small business and economic bases: Storey County's economy relies on revenues generated primarily by tourism, gaming and prostitution, while Lyon County's base is mainly farming, ranching and associated businesses. Lyon County also has a large area devoted to "bedroom communities" for Reno and Carson City, namely: Fernley, Dayton, Mark Twain Estates, Stagecoach, Silver Springs and Silver City. Each county is currently seeking to attract new "clean" business and industry to expand their respective revenue bases and diversify their economies.

The demand for additional services generated by the influx of residents who work in other counties has strained the budgets of both counties, yet these same residents vociferously oppose the establishment of certain types of business activities—such as mining, minerals processing, manufacturing, chemicals, or any other so-called "dirty" industry. They have even demanded that existing mining operations be closed down if they are visible from any highway, main county road, or residential community. This, they say, is to enhance the view shed by removing unsightly industrial operations.

And these demands are being made by residents of an area that is literally pock-marked by previous mining activities. They are, in effect, trying to impose big-city viewpoints on the populations of two rural counties. While working, shopping, banking and doing other business in either Carson City or Reno, they are residing in rural areas and, by their votes, are changing the economic and environmental conditions in those rural districts.

Foremost among the many anti-mining measures taken to date are: (1) zoning ordinances where patented and unpatented mining claims are zoned as residential areas; (2) the establishment of historical districts with the authority to issue or deny mining permits; (3) ordinances stipulating requirements of special-use permits for mining on both federal and private lands within the counties, with the authority to deny issuance of these permits; (4) organization of opposition groups in the residential communities, and (5) a media blitz by various environmental and residential groups to influence political and public opinion against mining interests.

Despite the overwhelming opposition resulting from these actions, several mining companies and individuals are still trying to open up new mining operations in the district. Why? Because economically viable mineral deposits have been discovered in the area, and the potential exists for more discoveries. This is logical, since the precious metals mineralization is very extensive and pervasive throughout the entire region. The early miners knew this, and it is the only reason that towns like Virginia City, Silver City, Gold Hill and Dayton presently exist.

It seems that by purportedly attempting to preserve Nevada's historical mining heritage, the preservationists are attempting to destroy the present and future mining heritage of the Comstock Lode district.

30
Nevada Politicians Target Mining for Tax Increase

CITING THE NUMEROUS PROBLEMS Nevada has with children's issues, education, crime, drug abuse, mental illness, health care, etc., and the need for pay increases for teachers and government employees, Nevada's Acting Governor Bob Miller recently posed this question in his State of the State speech before the Nevada State Legislature and the general public: "Where can we get the needed funding, without burdening homeowners and wage earners?"

His answer? "Mining in Nevada has never been more lucrative." Miller then proceeded to outline a wish list of benefits the state might obtain by exacting higher taxes from mining, and detailed his assessment of the industry itself. Needless to say, his opinions weren't complimentary. His impassioned appeal could be summarized in one statement: Nevada can fund most of its needed social programs and pay raises by imposing increased taxes on only one industry—mining.

Many other Nevada state legislators have targeted the state's mining industry for higher taxes and stringent state regulation, especially many assemblymen from Las Vegas. Las Vegas Assemblyman Marvin Sedway, chairman of the power Assembly Ways and Means Committee, has pursued a vendetta against mining for several years. Using this position to advance his attacks on the minerals industry. Sedway has advanced several burdensome mining tax proposals in the past—including his famous "Jim Dandy" gold "fee" bill in the 1987 Legislature that would have imposed a fee of $16.50 on every ounce of gold "extracted" in Nevada. Fortunately, Nevada only has a legislative session every other year.

Although his gold fee proposal was defeated in the Senate Taxation Committee in 1987, Sedway has publicly stated his intention to introduce other "more onerous" mining tax measures in the 1989 Legislature (one of which will have been introduced by the time this item is published). He has threatened that if his proposals fail in the Legislature, he will back a "public initiative" to increase mine taxes by a vote of the people in the state. Acting Gov. Miller has also said he will back a public initiative to raise mine taxes if his current proposals are not enacted by the 1989 Legislature.

Senate Joint Resolution 22 (S.J.R. 22, a proposal to amend the state constitution to increase the net proceeds of mines tax to a maximum of 5%; it is supported by the mining industry) was introduced and passed in the 1987 Legislature, but it must still be passed by the 1989 session and a vote of Nevada's residents. This measure would increase mining taxes by about 150%. According to Gov. Miller and Assemblyman Sedway, however, it is still not enough of an increase. Miller has presented an amendment to S.J.R. 22 so the measure will raise mining taxes by over 455% (over 4½ times the net proceeds of mines taxes paid under current law). Both Miller and Sedway maintain this increase is necessary to ensure that mining pays its "fair share" of state taxes.

Nevada prides itself on its favorable business tax structure—no corporate, business or income taxes—and is actively promoting this tax advantage to attract out-of-state and foreign capital investment in the state, along with economic diversification efforts directed toward getting businesses to relocate to Nevada. Only *two* businesses are taxed in the state: mining and gaming. Most other taxes are targeted so the major portion is paid by tourists visiting and gambling in Nevada. As a result, both gaming and mining are constantly being considered for higher taxes, and this is used to maintain the attractive tax climate for almost all other types of businesses. Other businesses, of course, are not required to pay their "fair share" of the state's taxes.

Acting Gov. Miller says his proposed amendments will only increase mining taxes to a level comparable to that found in most other states. In order to compute these "comparable" levels of taxation, Miller also included all other types of state taxes paid by mining firms in other states—income tax, severance tax, precious metal tax, mining license tax, indemnity tax, etc.—*none* of which currently exist in the state of Nevada.

In order to achieve his tax objectives, Miller has proposed an amendment to S.J.R. 22 (and a companion measure, S.B. 61) that would eliminate almost all of the current deductions allowed in the computation of net proceeds of mining operations. His amendment would *only* allow deductions for "the actual cost of extracting the mineral." No deductions would be allowed for exploration, drilling, development, ore reduction, milling, refining, transportation, sales costs, maintenance, repairs, equipment depreciation, insurance, unemployment compensation, royalties paid, management and office expenses, etc.

This would place the tax computation on a slightly reduced value of the gross production figure, as opposed to taxing the actual net proceeds (profits) of mines. Yet the term *net proceeds* would be retained, and the profits of mining operations would be reported on this basis—meaning the reported gross income and net profit figures would be only slightly different.

Passage of these proposals by the legislature in their current form would produce a tremendous impact upon mining activities in the state. Grassroots exploration for minerals would practically cease, low-grade ore deposits could not be mined, marginal mining operations would be forced to shut down, mining investments would drop dramatically, mine life for most operating mines would be severely reduced, and Nevada would slowly but surely lose its edge as the leading gold-producing state in the nation. Unfortunately, these negative impacts on the mining industry would be welcomed and applauded by many politicians and certain other people in the state.

31
Nevada Mine Tax Battle Grows Stronger

NEVADA GOVERNOR BOB MILLER and several Las Vegas legislators are constantly asking the people of the state and the Nevada State Legislature to choose between the children and the mining industry in order to gain acceptance of their plan to increase mining taxes by 455 percent. No other options are being presented or considered—the state's children can be saved only by extracting more money from mining.

Senate Minority Leader Joe Neal, D-North Las Vegas, recently delivered a passionate plea for the senate to support Gov. Miller's proposal to eliminate almost all operating costs from deductions allowed in computing the net profits of mines, a move that would have almost all the state's miners paying a 5% tax on 80 to 90% of their operating expenses and investment capital. He implied the senators would be rejecting children if they voted no on the governor's proposal.

"I would much rather stand with the children of this state than next to a carved-out mountain," Sen. Neal said. "We owe our children more than the leach fields of cyanide they use to extract gold from rocks."

Immediately following these statements, the senate rejected the proposal by a 12-to-8 margin.

"I will continue the fight," said Gov. Miller, who met with each individual senator before the vote was taken. "I feel we made real progress today toward assisting the children of our state. My resolve has not weakened. I will continue to push forward."

Gov. Miller launched his attack on the mining industry immediately following his ascension to the governorship, which resulted from the election of Governor Richard Bryan to the U.S. Senate. A few days after becoming acting governor, Miller publicly stated that he would go after mining to meet the state's needs for additional revenues. In his State of the State Address, he emphatically said there was only one choice: Nevada's children or mining.

These constant efforts to impose additional taxes on Nevada's miners are coming at a time when the mining industry and many state legislators have already agreed to increase mining taxes from 2% to 5% of net profits. Senate Joint Resolution 22, passed in the 1987 Legislature, calls for a 150% increase in state taxes on the net proceeds of mines, and it has been fully supported by the mining industry for two years. However, it is not enough for Gov. Miller and several Las Vegas politicians, who want to increase the amount agreed to by over three times. In order to achieve their objectives, they are insisting that the mine taxes be based on an amount close to the gross income received from production.

Senate Bill 61, a measure designed to implement S.J.R. 22, details the deductions allowed from gross income in the computation of the net proceeds subject to state taxation. The anti-mining faction is proposing that these deductions be disallowed, thereby raising the amount subject to a state tax by over three times the actual net profits of the mines. This action would force many of the marginal operations and smaller mines to shut down, because they could no longer operate at a profit. Repeated attempts to point out the unfairness of this tax increase and the resultant negative impacts have failed to reach the public, largely because the media refuses to publish the information—or even to look into the matter.

However, State Senator Charlie Joerg (R-Carson) has defended the original mining tax resolution (S.J.R. 22) on the senate floor and attempted to dismiss the charges raised by Gov. Miller that mining does not pay its fair share of state taxes.

"They pay taxes like every other citizen—sales taxes and property taxes," Joerg said. "They didn't escape taxation."

He said that when the state constitution was drafted in 1864, the framers decreed that mining would pay an additional tax on their net profit. This tax is based upon the value of the minerals produced minus legal deductions, and it is paid on top of all other state taxes paid by mining.

U.S. Rep. Barbara Vucanovich (R-NV) has also outraged Gov. Miller and some legislators with her remarks that lawmakers should refrain from overtaxing the mining industry.

"It will not help Nevada, or our education system, in the long run to disproportionately tax the mining industry because it is the course of least political resistance," Vucanovich said when she appeared before the Nevada Legislature recently. In a press conference afterward, she said she opposes Gov. Miller's proposals to raise an extra $66 million from the mining industry by removing their deductions. Along with several legislators and the mining industry, Vucanovich supports S.J.R. 22—which will raise an estimated $52 million in state revenues over the next two years.

Following their passage in the Senate, S.J.R. 22 and S.B. 61 moved to the Assembly Taxation Committee, where they came under further attack in testimony before the committee. Scott Craigie, Gov. Miller's chief of staff, surprised committee members by saying the governor will hold off on his proposal to extract another $66 million in mining taxes until after S.J.R. 22 and S.B. 61 are passed. He said that after Nevada's voters pass these measures in May, then the governor will push for an amendment to S.B. 61 that will remove most of the miners' deductions. In this manner, he said, the legislation can be voted in first and then changed at a later date.

"The governor believes very strongly this first step must be taken," Craigie said. "If it doesn't pass—a worst-case scenario—the mining industry would leave this session without any tax increase at all. That would be wrong. There is a second step, at least in our minds.

"I don't think anyone should think S.B. 61 freezes (mining's) taxes forever," Craigie said. He stated that Gov. Miller's plan will surface again in the form of a separate bill before the Legislature.

During a testy exchange between northern and southern Nevada legislators, Democrat Assemblyman Matt Callister of Las Vegas contended that the measure would extract far too little money from mining. He said that citizens are thoroughly confused by the state's mining tax policies, and the 750,000 people in southern Nevada will no longer stand for favoritism toward mining. He believes that mining taxes should be raised at any time the legislature finds a need for additional revenues.

"It is time we treat mining like other industries," Callister said. "We go from one bad decision, one half-ass decision to another."

The assemblyman failed to mention that only the mining and gaming industries pay any tax on income in Nevada, and that if mining was treated like other industries it would pay no taxes at all. He did say that mining should have to pay a tax similar to the 6% gross revenue tax paid by the gaming industry. He also promised to propose another constitutional amendment that would remove all references to mining taxes in the state constitution. Such an amendment would allow the legislature to routinely change mining taxes without having to worry about the constitution.

Assembly Minority Leader Lou Bergevin, R-Gardnerville, told the miners present that Callister's comments showed "Clark County is out to get you." He said later that about six Clark County assemblymen "screw the state anytime they can."

Following the debate over the issues, the Assembly Taxation Committee passed S.J.R. 22 but held on to S.B. 61 and assigned the bill to a subcommittee to discuss any proposed changes. It is expected that the mining tax battle will continue throughout the remainder of this session of the legislature and that it will not be resolved for many years. Regardless of this year's outcome, most politicians and miners expect that mining will be targeted for tax increases and other burdensome legislation every time the legislature convenes in the future.

And this is only one of the many mining bills introduced in the 1989 Nevada Legislature. Democrat Assemblyman Marvin Sedway of Las Vegas has been very busy in his pursuit of a vendetta against Nevada's miners. During his publicly announced "Bomb the Miners Week," Sedway introduced a mining bill every day—for a total of five new measures designed to exact a higher price in both fees and mining costs. A brief description of each of these mining bills is given below:

Assembly Joint Resolution 13: This measure proposes to amend the Nevada Constitution to allow state taxation of the gross proceeds of mines. This resolution would effectively cancel S.J.R. 22 and S.B. 61 and create a much more severe impact on the state's minerals industries than those described under Gov. Miller's plan. This is because the measure would allow no deductions of expenses from gross income before payment of state taxes.

Assembly Bill 172: This measure would charge a "fee" of $2 per troy ounce of gold and $.50 per troy ounce of silver extracted in Nevada, and these revenues would be used to create a program for the reclamation and rehabilitation of abandoned mine sites. Providing that all of the existing mines could continue to operate in the state at projected levels, industry representatives believe this measure would cost about $15 million annually.

In reality, several of the low-grade gold mining operations and almost all of the primary silver producers (those with no significant byproduct gold production) would have to shut down, thereby producing a significant reduction in the total quantity of these metals produced statewide. The $.50 fee per ounce of silver amounts to about 8.5% of the current billion price for silver, which would severely impact most silver producers.

Assembly Bill 178: This bill would establish a registry for information and impose a fee to accompany reports on the use of sodium and potassium cyanide compounds in Nevada. The fee amounts to $1 for every pound of cyanide compounds used in the state, and the industry estimates this would cost between $40 million and $45 million annually. It would more than double the average cost per pound for these chemicals.

This bill would create an impact ranging from major to severe, depending upon the size and profitability of each particular mining operation. Many of the lower-grade gold mining operations, primary silver producers and marginal mines would either be forced to shut down or cut back on operational expenses. The measure would definitely lower the annual amount of precious metals produced and shorten mine life.

Assembly Bill 190: This bill removes the right of eminent domain action for mining purposes, which allows the condemnation of property for mining purposes. It would have no significant economic or operational impact on the minerals industries, primarily because it is seldom used today. Mining companies currently lease or buy rights-of-way or properties directly from the owners. This measure is largely considered a harassment issue that is designed to detract from mining's importance in Nevada.

Assembly Bill 208: This measure provides for the verification of reported proceeds of mines and the numbering of gold and silver ingots that are produced in the state, as well as stamping them with the state seal. Called the "S.S. Bill" (after the Nazi Schutzstaffel, or storm troopers) by miners, this bill allows "unannounced inspections" of all buildings, equipment and workings at mining operations at any time for any reason without a search warrant. It implies that all miners cheat and steal, and the state has the right to exercise complete control over them. No other business in the United States is treated in such a degrading and demeaning manner. The bill also provides for penalties against anyone in the state who is caught transporting any precious metals that have not been stamped with the state seal and numbered in a state registry. Industry officials estimate it would take a minimum of 100 state agents to police and enforce the measure, as well as the cooperation of state law enforcement personnel.

In addition to the legislation mentioned above, there are also a number of pending bills that will deal with environmental concerns, mine reclamation and bonding at the state level. At least two state reclamation bills are expected to be introduced shortly, along with recommendations for increased personnel (and stringent new regulations) in the Nevada Department of Environmental Protection. The state's environmental groups are demanding that state reclamation requirements and bonding be placed under the NDEP for enforcement. Some of the more radical environmental activists are also engaging in attacks on mining at every opportunity.

As if this weren't enough, the governor and several politicians are actively spreading erroneous and misleading information about mining in the media, in an attempt to get more backing from the general public. In addition to the oft-repeated "children vs. mining" ploy, they are making inflammatory and demeaning statements about mining and miners. They stress the insulting implication that miners do not care for children and their education.

It turns out that Assemblyman Sedway's announcement was somewhat inaccurate: It is actually "Bomb the Miners *Year*" in Nevada.

After finally realizing that their very survival is at stake, Nevada's miners and prospectors have organized a grassroots movement to oppose the draconian anti-mining legislation being backed by the governor and several Las Vegas legislators. Since, for the most part, the state's major newspapers and television stations have been supporting the anti-mining politicians in their efforts, the miners are finding it difficult to get factual information before the general public and have been forced to adjust their efforts accordingly.

The miners have unleashed a flood of letters and telephone calls to all of the legislators, newspapers, TV stations and the governor, and these actions are slowly, but surely, counteracting the negative media bias and the erroneous and misleading information being put out by the governor, his staff, and the anti-mining legislators. As a direct result, some of these politicians are temporarily backing off and instituting several changes in their tactics.

The anti-mining faction is now promoting the original mining legislation that is supported by miners in an attempt to get the voters in the state to approve them, after which they will submit new legislation that will drastically alter the voter-approved measures to a form that will accomplish their original objectives. This blatant attempt to fool Nevada's voters has been outlined in public statements, most of which have not been accurately reported in the state's news media. So, the miners are being forced to take additional defensive measures against this new threat.

In the meantime, miners are continuing to present firm opposition to the many other anti-mining legislative proposals being considered by the Nevada Legislature—most particularly those bills authored by Assemblyman Sedway. These threatening measures are being addressed by the presentation of factual testimony in legislative hearings held on each of the proposed bills. Fortunately for the miners, Sedway is rapidly gaining a reputation statewide for his radical attacks on the mining industry, and the general public is beginning to show definite signs that they are doubting his credibility. Although there is still a long way to go in combating his damaging proposals, it is generally felt that Sedway has a very poor chance of getting his bills through the legislature this year.

In one way, these major attacks on Nevada's mining industry this year are serving a useful purpose: they are acting to galvanize miners into an effective force in the political process. Regardless of the eventual outcome of this year's battle, the state's miners have realized that they must take a stand together or they will not survive.

32
Nevada Mine Tax Increase Approved by Voters

NEVADA'S VOTERS SHOWED overwhelming support for a mining industry-backed initiative to increase mine taxes on the May 2, 1989, ballot. By passing Question 1, voters changed the Nevada Constitution to allow the tax on mining's net proceeds to increase from about 2% up to a maximum of 5 percent. The state's counties will retain the original 2% tax and the balance, estimated at roughly $52 million for this biennium, will go to the state.

Nevada Mining Association President Jim Gourdie said the industry has honored a commitment it made during the 1987 Legislature by endorsing and campaigning for passage of the mine tax measure. He said that approval of the proposition by voters means the mining industry is now paying its "fair share" of state taxes. The May 2nd vote shows an awareness of the people that this is a fair tax on mining in the state, he added.

Even in rural Nevada, location of almost all of the largest gold mines, voters decisively backed the proposition to change the 125-year-old state constitution and increase the tax on the net proceeds of mines. When all of the votes had been counted, Question 1 had received approval by 107,679 voters (or 77.8%), while the remaining 30,663 voters (or 22.2%) had cast their ballots against the proposition. Nevadans had rejected a similar mine tax measure just five years earlier.

However, during this general election the governor, miners, educators and legislators nearly fell over each other in backing the mine tax measure. In order to fend off attempts in the current legislature to tax mines at an even higher rate, mining companies and small miners pledged to support the constitutional amendment to increase the net proceeds of mines tax. The minerals industry even put $150,000 into an advertising campaign to sell the tax increase to the state's voters. In spite of the widespread publicity the measure received only about 36% of the state's registered voters turned out to cast ballots during this year's election.

The voters' more than 3-to-1 support for Question 1 meant that Gov. Bob Miller's 1989-91 state budget plan remains basically intact and fiscal chaos has been avoided. The governor had already incorporated the estimated $52 million in increased mining revenues into his proposed $1.6 billion operating budget for the biennium. If the voters had not approved the tax measure, a gaping hole would have been blown in the budget and the Nevada Legislature would have had to find additional sources of revenues to make up the shortfall.

Earlier this year, Gov. Miller had backed plans to exact another $66 million from the miners by eliminating all of their allowable tax deductions except one—the actual cost of extracting the minerals. Miller's team unsuccessfully promoted that plan in February before the Senate Taxation Committee and then, refusing to accept defeat, placed the plan before the full senate. Failing again in obtaining enough support in the legislature, he decided to back the original tax measure—at least until it was passed by the voters. His chief of staff, Scott Craigie, said the governor would attempt to eliminate the deductions from the calculation of the net proceeds of mines sometime after the election. However, he recently said he would talk to legislative leaders from both parties before making a move.

But a recent news report states that Assemblyman Gary Sheerin, D-Carson City, is expected to introduce a bill to reduce the deductions that mining companies may take before they compute their taxes, which is almost the same as the governor's proposal. The reduction or elimination of most actual operating costs would increase the amount of income subject to taxation, thereby making the taxable revenues much nearer to gross income. This type of legislation would increase state taxes as effectively as if the percentage rate of taxation was increased.

Senate Joint Resolution 22 (the measure passed by the voters) just amends the constitution to allow mine taxes to be assessed at a maximum rate of 5%. However, Senate Bill 61 (also passed by the Nevada Legislature) outlines the means by which S.J.R. 22 will be implemented. S.B. 61 stipulates the means by which the taxes will be paid, defines other relevant procedures, and enumerates the allowable deductions from gross revenues to calculate the actual net proceeds of mines. The bill also contains a percentage-ratio table of net proceeds to gross proceeds that sets the actual percentage rate of taxation which individual mines must pay. The less profitable mines will pay state taxes at a much lower rate (close to 2%) than the tax rate paid by the most profitable mines (close to 5%). This equitable scale of taxation ensures that marginal mines will not be subjected to an unreasonable tax burden, while allowing a higher tax on the state's most profitable mines—which can best afford to pay it.

However, the current boom in Nevada gold mining has caught the eye of several state politicians who see an opportunity to obtain more funding for the expansion of state programs and increased revenues for raising the pay of state employees and teachers. As a result, S.B. 61 has become the prime target for "enhancing" state tax revenues. Despite the fact that only mining and gaming pay any form of income tax in Nevada, these politicians maintain that mining "owes" the state more money because the major gold companies are, for the present at least, extremely profitable. They say this is not a biased form of taxation because the gold "belongs to the people of the state" and it is a nonrenewable resource. This group of legislators and other politicians have been joined by the anti-mining groups in efforts to further tax and control the state's minerals industries.

As a result, numerous burdensome mining measures have been introduced in the 1989 Nevada Legislature (see "Nevada Mine Tax Battle Grows Stronger" in the April 1989 issue of the *CMJ* [immediately preceding article in this volume of the Dave W. Parkhurst Mining Writing Collection, article 31]). Several of these bills would inflict serious damage on Nevada's mining industry. Now that the initial mine tax measure has been passed by the voters, it is expected that these anti-mining factions will increase their efforts to enact legislation that would seriously impact the state's mining community. Most of these measures have already been introduced in the legislature, but they were placed on a temporary hold until after the election. So, miners will have to continue to present firm opposition to these potentially damaging proposals.

33
Nevada Legislature Seeks More Mine Tax Revenues

FOLLOWING THE APPROVAL of a major mining tax increase by Nevada's voters in early May (see "Nevada Mine Tax Increase Approved by Voters," June 1989 *CMJ* [article 32 in this volume]), several of the state's leading politicians and legislators launched an expected campaign to obtain additional tax revenues from the mining industry. Several more mining tax bills and related proposals surfaced in the latter part of May, and hearings were scheduled on some existing tax measures that had been placed on a temporary hold until the main mining tax increase had been approved in the May election.

Assemblyman Gary Sheerin (D-Carson City) introduced Assembly Bill 770, which would create a new tax schedule that would require mines with profits of more than $120,000 per year to pay the maximum 5% state tax rate. A.B. 770 also attempts to eliminate the deduction of many actual mining costs in the computation of the net proceeds of mines. As a result, it would increase the amount subject to state taxes by adding capital costs to the real net profit from mining operations. The proposed revision in the tax scale, coupled with the increased amount of "net" proceeds, would automatically elevate almost all of Nevada's mines into the top tax brackets.

Assemblyman Matt Callister successfully scheduled three hearings on Assembly Joint Resolution 21, a measure designed to nullify the intent of the mining tax measure just passed by the state's voters, which would make mining constantly vulnerable to additional taxation in every session of the legislature. During the first hearing on the proposal, Nevada Mining Association President Jim Gourdie told the Assembly Taxation Committee that "it would appear that the goal of A.J.R. 21 is to enable the state, at will, to tax the same asset in multiple ways. To do so would scrap the long-evolved system, which has successfully allowed mining to grow in this state while paying a reasonable rate of taxation."

A representative from the Nevada Miners and Prospectors Association said, "We have been asked to place our trust in the fairness of the legislative process, yet we are being inundated by numerous burdensome proposals currently being considered in this session of the Nevada State Legislature. Our perception is that we, and our means of making a living, are under attack."

Governor Bob Miller has also jumped into the mine tax battle again, proposing that a school class-size reduction should be funded by increasing mine taxes another $16 million. Miller has been privately asking legislative leaders to support a plan to cap deductions used for calculation of the net proceeds of mines. His proposed caps on deductions would include $1 million on depreciation, $2 million on processing, $250,000 on transportation, $100,000 on insurance, and $40,000 on marketing. Miller's proposals have not as yet garnered much legislative support.

However, the Assembly Committee on Ways and Means recently introduced Assembly Bill 866, which would include legislation to accomplish much the same purpose as the governor's proposals. A.B. 866 stipulates that the rate of tax upon any operation for which the net proceeds in a calendar year exceed $4 million is 5 percent—the maximum rate under current law. The bill also proposes a 10% penalty for any mine that is not within 90% of the estimated net proceeds (paid in advance) for any year, with the penalty applying to the full amount of taxes paid as opposed to the actual amount of taxes underpaid. In addition, miners would be required to file quarterly reports on the exact amount of their gross and net proceeds.

At the first hearing on A.B. 866 in early June, miners pointed out the fact that it would be almost impossible to make an accurate estimate of each year's net proceeds in advance, largely because of fluctuating gold and other mineral commodity prices. They also said a mine that processes higher-grade ores or increases production would be penalized for increasing revenues. They added that the measure was just another attempt to impose stringent and onerous controls on Nevada's minerals industries.

At the first hearing in early June, both mining industry spokesmen and county representatives told the Assembly Taxation Committee they were flatly opposed to Sheerin's A.B. 770. Miners added that the bill was just another one of the numerous proposals being considered that were designed to impose the maximum possible tax on the state's minerals producers. They said A.B. 770 would even require break-even and marginal producers to pay much higher taxes—regardless of whether or not they are actually making a profit. Some producers would pay taxes on investment capital under the bill, a condition that miners said would be intolerable.

In mid-to-late May, Nevada's miners were once again being trashed by several politicians and legislators, as well as some newspapers. They accused the mining industry of withholding information from the voters in order to get Question 1 passed (the main mine tax measure), expressing outrage that the sliding scale of taxation was suddenly "discovered" after the election. However, the same sliding scale had been referred to numerous times in the press for the past few months, and several legislators had also tried to change the ratio in order to obtain higher tax revenues (also widely published). Miners pointed out that a potential projected shortfall in mining taxes was entirely due to the drop in gold prices, a fact they had pointed out numerous times since the first of the year.

In addition to the proposed legislation mentioned above, there are a number of other mining bills advanced in this session of the Nevada Legislature that have not, as yet, been resolved. All of Assemblyman Marvin Sedway's (D-Las Vegas) self-proclaimed "Bomb the Miners" bills are still sitting in various assembly committees (a total of five anti-mining bills), and they could surface at any time in the waning hours of this legislative session. Sedway's vendetta against the state's mining industry is well-known, and it is not likely that he would miss any opportunity available to pass at least some of his proposed legislation.

Some current estimates place the end of this year's legislature from mid-June to early July, but many important issues have not yet been resolved. Several general state tax increases are still being discussed, and the general view is that at least some form of additional taxes will still be required to meet the state's proposed budget for the biennium. As a result, it is also expected that many more attempts will be made to pass additional mining taxes, because it seems the course of least resistance. So, miners will have to continue their vigilance and opposition to the remaining anti-mining measures.

34
Nevada Mine Tax Issue Finally "Clarified"

NEVADA'S MINERS BREATHED a collective sigh of relief as the Nevada State Legislature finally wound up the 1989 session and went home until (hopefully) January 1991. Finishing the longest session in the state's history in the early morning hours of July 1, legislators had also racked up a new record in the number of potentially damaging mining bills proposed and considered. Fortunately, cooler heads prevailed in the frantic rush to enact as much legislation as possible before adjournment, and none of the anti-mining measures saw the light of day.

In the final analysis, not one of Assemblyman Marvin Sedway's five "Bomb the Miners" bills received serious consideration. Assemblyman Matt Callister's efforts to remove constitutional protection from the state's miners did not fly very well either, despite the fact that he succeeded in holding three hearings on A.J.R. 21 before it was killed. Acting Governor Bob Miller's efforts to eliminate mining deductions to produce a massive increase in mining taxes also went by the wayside, and Assemblyman Gary Sheerin's attempt to eliminate major deductions and revise the tax table were gutted from his mine tax measure before it received approval.

However, the legislature's final push to at least increase the state mine tax revenues was moderately successful. After removing proposals to eliminate tax deductions, revise the state tax table, provide for onerous penalties for underestimating tax payments, and require stringent quarterly reporting requirements, legislators finally passed a new and improved bill (A.B. 770) that capped the net proceeds rate at $4 million. The net result requires miners to pay the full 5% state tax rate on all of their net proceeds if the amount exceeds $4 million. The authorized deductions from gross proceeds in the computation of net proceeds were left intact.

Except for the $4 million net proceeds cap, the original S.J.R. 22 and S.B. 61 approved by the state's voters in May emerged practically unscathed. Since these original measures have been fully supported by Nevada's miners for the last two years, this resolution of the mine tax issue is being viewed as somewhat of a victory for the miners.

However, the hard-fought battles for fair and reasonable treatment of miners and the anger aroused by the frequent demeaning and derogatory statements made by certain politicians and legislators will not be easily forgotten. Miners have become fed up with the constant trashing of mining by public officials and the news media, and it is very likely that this resentment will manifest itself in future elections. After all, how often do miners and the industry as a whole get publicly trashed in a State of-the-State address by a new governor?

It is also unlikely that miners will forget the sponsors of the numerous anti-mining measures proposed this year, or their incessant public attacks on the industry. These actions would seem to indicate just what mining interests can expect in the next session of the Nevada Legislature.

PART V

Metals and Minerals Mining in the Silver State

Plaques displaying the Governor's official proclamation were presented to officials of the Nevada Mining Association and Nevada Miners and Prospectors Association at a ceremony held in the governor's office on March 2, 1990....

"In recognition of Nevada's mining industry, which created, established and maintained our state's industrial cornerstone, resulting in benefits to the Nation and the World."

35
Nevada Enacts Mine Reclamation Law

FOLLOWING SEVERAL MONTHS of extensive debate, revisions and additions, the Nevada State Legislature approved the state's first mined lands reclamation laws in its history. Assembly Bill 958 was passed by both the senate and assembly on June 24 and was signed into law by Governor Bob Miller a few days later.

Senator Charlie Joerg (R-Carson) said, "This is the first mining reclamation bill in the state of Nevada in 125 years. I think, obviously, the time has come for this."

"It's the absolute first hardrock mining bill passed in the state," echoed Assemblyman Ernie Adler (D-Carson), following the assembly's concurrence in several senate additions and changes to the bill.

The compromise bill embodied several major provisions contained in two bills introduced during March, A.B. 315 from the assembly and S.B. 212 from the senate. Both of these bills underwent extensive revision following several hearings on the issue, only to be killed as the legislative session entered the last few days before adjournment. However, both bills were resurrected from the dead in the waning hours of the session and then died again after they were replaced by a new proposal that combined elements of both bills.

A.B. 958 was introduced during the last hectic hours of the session and given top priority by both the assembly and the senate. As with the previous two bills, A.B. 958 underwent extensive revision and amendment. In this case, however, the necessary changes were made over a period of just a few days. Noticed hearings on the measure went by the wayside, and most of the bill's content was frantically arranged and written in the hallways of the legislature. No formal draft of the bill was available for public inspection until the morning of the day that the bill was passed.

As a result, representatives of the state's two major mining groups, the Nevada Mining Association and the Nevada Miners and Prospectors Association, found themselves in the position of having to track down copies of the latest bill draft requests (BDRs) in the hallways in order to get some idea of what was being proposed. Because of this situation, most of the miners' input on the bill was hurriedly arranged during the last few hours before the bill was passed.

Because of the haste and confusion (several major modifications and additions were not even available in printed form), several errors were left in A.B. 958 when it received final approval. This caused even more confusion until it was decided that a "trailer" bill could be drafted that would amend A.B. 958 and make the necessary corrections to the measure.

Senate Bill 549 was hastily devised, printed overnight and approved in a short period of time. However, the initial form of S.B. 549 did not address all of the modifications required, and it also had to be hastily amended. The final version of this trailer bill came out a few days after the original version was approved, and this last revision to A.B. 958 was finally passed just before the session ended.

In its final form, A.B. 958 placed the responsibility for enforcement in the state Division of Environmental Protection and created a funding mechanism whereby the state's mining interests would be charged fees to support the agency's costs in administering the reclamation program. In addition, another funding mechanism was created so the state's miners would pay the costs of administering an abandoned mines hazard abatement program, which will be the responsibility of the state Department of Minerals. So, A.B. 958 requires that the mining industry pay for all of the state's costs associated with administering and enforcing the reclamation laws.

The State Environmental Commission will be responsible for adopting the regulations necessary to implement the reclamation program, and one new member with mining reclamation experience will be added to the commission. The Department of Minerals will retain some oversight of the reclamation conditions placed on mining plans of operations and exploration permits by the Division of Environmental Protection.

The new law provides that federal approval of mining plans of operations and exploration projects supersedes the state requirements if the federal agency requires adequate reclamation and bonding to meet state requirements. It also provides that the Division of Environmental Protection will enter into a memorandum of understanding with the U.S. Forest Service and the BLM concerning administration of reclamation requirements on public and private lands.

Small mining operations and exploration projects of under 5 acres in surface disturbance will not be required to post reclamation bonds, unless the mining operation removes over 36,500 tons of materials in any calendar year. Both small mines and exploration projects will, however, be required to file reclamation plans and be responsible for complete reclamation of surface disturbances.

The major provisions of the state reclamation laws will take effect October 1, 1990, while the framework for adopting state regulations on mined-land reclamation and the abatement of hazardous conditions at abandoned mines take effect immediately. The bill amends Title 46 of the Nevada Revised Statutes.

36
Nevada Legislative Actions on Mining Issues in 1989

ACCORDING TO A RECENT Nevada Mining Association newsletter, the state's industry managed to get through the 1989 legislative session in pretty fair shape despite the numerous anti-mining measures introduced this year. The NMA provided the following summary of the various mining-related proposals considered by the Nevada State Legislature.

S.B. 7 — Authorizes the Department of General Services to contract for the minting of precious metal medallions and bars; limits permissible use of the state's seal. Passed and signed by the Governor on 3/29/89.

S.B. 61 — Implements legislative matters in S.J.R. 22 of the 1987 session. Sets up an indexing schedule for payment of net proceeds of mines taxes; lists allowable deductions; provides for reports and payments of estimated tax (see S.J.R. 22 and S.B. 770). Passed and signed by the Governor on 3/20/89.

S.B. 212 — Provided for reclamation of mined lands in the state. Died in the Assembly Natural Resources Committee.

S.B. 287 — Requires payment of certain fees directly to the Department of Minerals. Passed and signed by the Governor on 7/2/89.

S.B. 469 — Eliminates the liability for fencing dangerous conditions at abandoned mine sites under certain conditions. Passed and signed by the Governor on 7/2/89.

S.B. 476 — Makes various changes relating to the protection of wildlife in the state. Basically implements practices previously agreed to by the mining industry and the Department of Wildlife. Passed and signed on 6/28/89.

S.B. 549 — Makes technical correction to A.B. 958 on reclamation (see A.B.958). Passed and signed on 7/5/89.

S.J.R. 22 — Constitutional amendment establishing a separate tax on the net proceeds of minerals; allows net proceeds tax to rise to maximum of 5% (see S.B. 61 and A.B. 770). Passed and submitted to voters as Question 1 on ballot. Passed by voters on May 2.

A.B. 65 — Requires Governor to appoint a person with experience in mining taxation to the Nevada Tax Commission. Passed and signed on 5/16/89.

A.B. 155—Provides for various changes concerning property taxes. Passed and signed by the Governor on 7/5/89.

A.B. 172 (Sedway)—Created a program for reclamation of abandoned mine sites by charging a per-ounce fee on gold and silver. Died in Assembly Natural Resources Committee.

A.B. 178 (Sedway)—Established a registry for information and imposed a fee and reports on the use of sodium and potassium cyanide. Died in Assembly Natural Resources Committee.

A.B. 190 (Sedway)—Removed mining as a public purpose for which right of eminent domain may be exercised. Died in Assembly Judiciary Committee.

A.B. 208 (Sedway)—Provided for verification of reported proceeds of mines and numbering of gold and silver ingots; surprise inspections of all mine facilities. Died in Assembly Government Affairs Committee.

A.B. 315—Required reclamation of lands subjected to mining operations and exploration projects. Died in Senate Taxation Committee.

A.B. 770 (Sheerin)—Revises provisions governing calculation and distribution of net proceeds of mines tax and sets up a trust fund for excess net proceeds revenues; provides that any mining company with $4 million or more in net proceeds will pay tax at a maximum 5% rate; outlines penalties for failure to estimate net proceeds within 90% accuracy; allows for estimate of net proceeds in each quarter to avoid penalties. Passed and signed by the Governor on 7/2/89.

A.B. 801—Revises provisions governing maximum revenues local governments are allowed to receive from certain taxes; provides for interim study of taxation methods in the state. Passed and signed on 7/6/89.

A.B. 815—Authorizes refund of money lost by counties because of credits for tax paid on net proceeds of mines. Passed and signed on 7/2/89.

A.B. 850—Prohibited disclosure of information from the IRS concerning the net proceeds of mines. Died in Assembly Taxation Committee.

A.B. 866—Made various changes related to the taxation of net proceeds of minerals. Died in Assembly Taxation Committee.

A.B. 958—Requires reclamation of lands upon which mining or exploration projects are conducted; provides fees for funding program administration; provides fees for funding abandoned mines program; provides funds for mapping service by Nevada Bureau of Mines. Passed and signed on 6/28/89.

A.B. 961—Authorizes and provides funding for certain capital improvement projects. Passed and signed by the Governor 6/30/89.

A.I.P. 1—Corporate income tax initiative by teachers to create a 10% tax on corporate net profits; requires use of proceeds for education. No action taken by Legislature, so initiative will appear on the 1990 ballot.

A.J.R. 13 (Sedway)—Proposed amendment to Nevada Constitution to change net proceeds of mines to gross proceeds for purposes of taxation. Died in Assembly Taxation Committee.

A.J.R. 21 (Callister)—Proposed constitutional amendment to eliminate mine tax provisions in Nevada Constitution. Died in Assembly Taxation Committee.

The NMA newsletter noted that Assemblyman Marvin Sedway's "Bomb the Miners Week" gained none of the widespread support that Sedway had claimed. Only one bill was heard in committee (A.B. 172) and it promptly died thereafter. The rest of Sedway's bills were not even heard.

The passage of Question 1 (S.J.R. 22) by both the Nevada State Legislature and a vote of the people of the state seemed to reduce any appetite for additional taxes on the state's minerals industry, and it also served to reduce the attractiveness of almost all the anti-mining proposals.

37
Governor Honors Nevada Mining Industry

NEVADA GOVERNOR BOB MILLER declared the week of March 12, 1990, as "Mining Industry Appreciation Week," in recognition of the mining industry's record gold and silver production last year. According to figures compiled by the Nevada Department of Minerals, gold production surpassed 5 million troy ounces and silver production was well over 19 million troy ounces for 1989. Nevada leads the nation in the production of both metals.

Gov. Miller said, "The 1989 production figures indicate the vitality of the state's second largest industry. The companies and their employees who have made the results possible are to be congratulated."

The 1989 production totals topped the 1989 figures of 3,676,000 troy ounces of gold and 19,535,000 troy ounces of silver. The 36 percent increase in gold production is attributed to major expansions at existing mining operations and to several new mines that came on line during 1989.

Plaques displaying the governor's official proclamation were presented to officials of the Nevada Mining Association and Nevada Miners and Prospectors Association at a ceremony held in the governor's office on March 2, 1990. Excerpts from the proclamation follow:

"*Whereas*, the minerals extracted from the earth have opened doors to progress throughout history and are vital to the continuation of our civilization; and

"*Whereas*, a productive and efficient mineral industry is basic to the economic health and strategic well-being of Nevada and the Nation; and

"*Whereas*, mining, our state's second largest industry, is a cornerstone of Nevada's economic structure and mineral products are essential to our way of life and our technologically-advanced society; and

"*Whereas*, the mining industry in Nevada has played a vitally important role in the establishment and growth of this state and, today, continues to experience dynamic change due to advancing technology; and

"*Whereas*, the mining industry in Nevada has reached a new pinnacle in gold production in 1989 of 5,000,000 ounces; and

"*Whereas*, the mining industry in Nevada also continues to lead the Nation in the production of both gold and silver; and

"*Whereas*, revenues produced by mining production and employment are a primary source of income for our schools, development of our infrastructure, support for our local governments; and

"*Whereas*, it is important for all Nevadans to understand the importance of mining and the characteristics of the mining industry responsible for providing our society with mineral products;

"*Now, therefore, I, Bob Miller, Governor of the State of Nevada,* do hereby proclaim the week of March 12, 1990, as a week to recognize and appreciate Nevada's mining industry, 'An Integral Part of Our History—A Partner in Nevada's Future';

"In recognition of Nevada's mining industry, which created, established and maintained our state's industrial cornerstone, resulting in benefits to the Nation and the World."

38
Small-scale Mining in Nevada

SMALL-SCALE MINERS AND PROSPECTORS were the first to discover and develop almost all of Nevada's major mining districts, and they still continue to provide most of the newly discovered mineral exploration and development properties today. It has been estimated that small-scale miners have been responsible for the discovery of over 90% of Nevada's producing mines, both past and present.

This contribution by small-scale miners is widely recognized and valued by most of the larger mining companies in the state. In fact, a large number of the state's major producing mines, which provide the bulk of Nevada's minerals, production, are actually owned by small-scale miners. These mining properties have been leased or sold to larger mining firms because they have the necessary infrastructure and investment capital to explore, develop and mine them, but the underlying property owner is receiving lease payments and/or mineral production royalties from the mine operators. A few contract (or grubstake) prospectors also retain an interest in some of the major producing mines and are receiving lease/royalty payments.

For example, individual claimholders or small partnerships are now receiving substantial payments from the following major producing mines: Jerritt Canyon, Paradise Peak, Chimney Creek, Rain, Marigold, Pinson, Dee, Florida Canyon, Crofoot, Rochester, Bootstrap, Lewis, Maggie Creek, Mother Lode, Preble, Sterling, Ratto Canyon, Fondaway Canyon, and others. Some of the major mining companies making payments to underlying claimholders include: Independence Mining Co. (formerly Freeport Gold), FMC Corp., Newmont Mining Co., Goldfields Consolidated, Rayrock Yellowknife, Pegasus, Hycroft, Coeur d'Alene, Tenneco, and others.

An even larger number of mining firms are making lease or advance royalty payments to small-scale miners under agreements covering exploration and development properties has been of mutual benefit to both, and this is continuing to provide much of the basic foundation upon which Nevada's mineral production depends.

In addition to discovering, claiming and leasing mineral properties to larger mining firms, quite a few of the state's small-scale miners are still running their own mining operations. Most of these mines, however, are producing minerals other than precious metals.

For purposes of classification, "small" mining operations in Nevada are generally considered to be those that employ 20 or fewer workers. All mining operations in Nevada are required to report their annual production to the Nevada Department of Minerals, and this data can be broken down to indicate small, medium and large mines according to the relative size of the mining operations. According to the most recent figures available (for 1989), small mining operations in the state produce the following minerals: aggregate, sand and gravel, clay, barite, gold, silver, antimony, zeolite, diatomite, dolomite, fluorspar, cinder, mercury, opals, limestone, gypsum, salt, landscape rock, talc and perlite. The bulk of Nevada's small mine production comes from industrial mineral deposits.

Some of the small metal and nonmetal mines, however, produce significant quantities of minerals with very few employees. For example, the Boss Mine in Esmeralda County reported a 1989 production of 8,983 ounces of gold and 6,948 ounces of silver with 2 employees. The McLean Mine in Esmeralda County produced 700 ounces of gold and 1,000 ounces of silver with 1 employee. The Dusty Mine in Pershing County produced 58 ounces of gold with 8 employees. The Trinity Silver Mine in Pershing County produced 718,714 ounces of silver and 70 ounces of gold with 7 employees.

Production of minerals other than precious metals with relatively few employees included: the Royal Peacock Opal Mine in Humboldt County, which produced 35 ounces of opals with 1 employee; the Clipper Mine in Lander County, which produced 15,172 tons of barite with 5 employees; the Crowell Fluorspar Mine in Nye County, 900 tons of fluorspar with 1 employee; the P & S Barite Mine in Nye County, 854 tons of barite concentrate with 4 employees; the Choats Sunrise Mine in Washoe County, 3 tons of antimony stirnite with 3 employees; the Ash Meadows Plant in Nye County, 1,000 tons of zeolite with 4 employees, and the Blanco Mine in Esmeralda County, which produced 726 tons of clays with 6 employees.

In other mineral materials, the Cind-r-Lite Co. in Nye County produced 53,948,000 pounds of cinders with 2 employees. There are also a large number of small sand and gravel operations that have a high production with very few employees.

Small precious metal placer operations in the state during 1989 included the Gold Bug Placer in Lyon County (reporting 211 ounces of gold and 1,081 ounces of silver) with 12 employees and the Dayton Sand & Gravel Co. in Lyon County (reporting 296,480 tons of sand and gravel and 151 ounces of byproduct gold) with 17 employees.

Some of the other smaller precious metals mines included: the Haywood-Santiago Mine in Lyon County, which produced 1,726 ounces of gold and 25,016 ounces of silver with 11 employees; the Illipah Mine in White Pine County, 3,879 ounces of gold with 10 employees; Project Glister in Eureka County, 8,451 ounces of gold and 23,520 ounces of silver with 12 employees, and the Little Bald Mountain Mine in White Pine County which produced 5,500 ounces of gold and 1,500 ounces of silver with 19 employees.

Overall, 81 smaller mining operations with 20 or fewer employees filed production reports with the state, only 18 of which were mentioned in the preceding examples of small-scale mines in Nevada. In addition, a significant number of new small placer and lode mines are currently being developed in the state, and these operations will be in production in the near future. Some existing small mines are presently stockpiling ores for eventual processing in custom mills and/or for future heap-leaching operations. A few older mines are also being refurbished prior to the resumption of mining operations.

Most of Nevada's small-scale miners and prospectors are still actively engaged in exploration for new mineral deposits. The lion's share of the grass-roots mineral exploration in the state is still being performed by individual prospectors. Open-pit, massive ore bodies are the primary objective being sought because these deposits represent the best economic potential, but all mineral discoveries, regardless of size, are being carefully investigated for potential mineral development.

Nevada's mineralogical, geological and political environment currently represents an excellent opportunity for miners, both small and large. The state's mineral potential for both hardrock and industrial minerals is still among the best in the nation. Barring any major shift in national mineral policies, Nevada's mineral industry is poised for rapid growth and a bright future.

39
Nevada Miners and NDEP Working on Regulatory Changes

THE NEVADA STATE SENATE Committee on Natural Resources' special mining subcommittee held several hearings on the problems that the state's miners were having with certain Nevada statutes and regulations during April and May, at which time the main problem areas and issues were defined. The senate subcommittee requested a report from the Nevada Division of Environmental Protection that would specifically address all the action taken to alleviate problems with the agency's enforcement and administrative policies. Subcommittee Chairman Sen. Ray Shaffer also requested a condensed list of specific problems from the small-scale miners and from the large mining companies as well. The miners' list was delivered to Sen. Shaffer on April 24, and the NDEP's list was submitted by April 30.

During this same time period, representatives from the small miners association and the large mining companies met personally with the NDEP's director, Lew Dodgion, on April 19 to discuss problem areas as well as proposed solutions. Agreement on a number of issues was reached, including a major shift in the NDEP policy toward a helpful role in regulating the mining industry, simplification of required forms and checklists, procedures for clarifying fee and regulatory requirements, and the setting up of regular meetings with miners to discuss changes in certain regulations. A number of these points were subsequently outlined in the NDEP report to the Senate Natural Resources Committee.

The full Senate Natural Resources Committee held another hearing on these issues on May 6, at which time the NDEP delivered their report and miners presented their lists of problem issues and proposed solutions. As a result, many of the major problems being experienced with the NDEP's policies in the past were satisfactorily resolved, and provisions were made to initiate necessary regulatory changes in the near future.

The ongoing program to address regulatory problems and provide corrective actions that resulted from these legislative hearings was successfully launched in July, with representatives from the state's two major mining associations meeting with NDEP Director Dodgion and his staff to initiate the process. As agreed to in the April meetings with Dodgion, the agency had already simplified the informational filing requirements for small mining operations under NRS 519A (state reclamation statute) and redefined the "small-scale facility" definition in NAC 445 (water pollution control regulations), which now reads: "...which chemically processes less than 36,500 tons of ore per year and no more than 120,000 tons of ore for the life of the project at any one permitted site." This revision was approved by the Nevada State Environmental Commission in Reno on July 31.

During the same Commission meeting, the NDEP amended the reclamation regulations dealing with drill holes to compliance with the Division of Water Resources Well Drilling Regulations (NAC 534), thereby eliminating the necessity for bonding down-hole abandonment and only requiring bonding for the surface cement plug. The change also allowed for miners to notify the NDEP in regard to any drill holes that would be mined out and eliminated any bonding requirement for this type of drill hole.

The NDEP's new policy directive is to work with the mining industry in a helpful manner, and the agency is also currently working on several other problem areas in specific regulations that were discussed during the July meetings. For example, Dodgion says the NDEP inspectors will no longer cite miners for regulatory violations on the first visit to a site, but they will instead notify the responsible party of the violation and necessary corrective action and will only issue citations if the violation remains at the time of a follow-up inspection. In addition, the agency is currently considering the implementation of a grievance procedure, wherein individual miners or mining firms that are experiencing problems with the agency can meet with the NDEP director and appropriate staff to iron out disagreements, rather than having to register appeals with the State Environmental Commission for problem resolution (which is the current regulatory procedure).

Overall, this program appears to be the most effective accomplishment of the state's miners that resulted directly from actions taken during the 1991 session of the Nevada State Legislature. This process is working out very well, and it has already resulted in positive actions being taken to alleviate a number of regulatory problems and in major changes being made in the state's administrative and enforcement policies. According to reports from both individual miners and mining firms who have dealt with the Nevada Division of Environmental Protection recently, the new policies and procedures are already working very well and most miners are generally pleased with the program.

40
Nevada Will Fight Unreasonable and Unenforceable EPA Standards

THE FEDERAL ENVIRONMENTAL PROTECTION Agency is now poised to impose excessively stringent water standards nationwide, an action that will cost industrial facilities and sewage treatment plants in Nevada and most other states millions (possibly billions overall) of dollars by early next year. However, Nevada officials have announced that they plan to fight the EPA's new "toxic" chemical standards for wastewater, because they say the strict requirements are both unreasonable and unenforceable.

"The economic impact is ridiculous and the improvement is not noticeable," said Dick Reavis of the Nevada Division of Environmental Protection (NDEP) recently.

In the biggest enforcement action ever, under provisions in the 1965 Clean Water Act, the Federal EPA announced on November 6, 1991, that the agency would impose its own standards for 105 so-called toxic chemicals on those states which had failed to act upon a 1987 congressional mandate to adopt the stringent standards themselves. The new limits will apply to chemicals present in all treated wastewater discharges.

Reavis said Nevada has its own set of toxic chemical standards for wastewater discharges, but the list is limited to the 60 or so toxic chemicals that scientific research has proven to be harmful. He said the state's standards are based upon each particular chemical's toxicity to human and aquatic life. By contrast, the Federal EPA standards list numerous additional substances that have not been proven to be dangerous, and many of the chemical limits are so low that they cannot be accurately measured with the current technology available.

Reavis added that the states which adopted the federal standards on their own likely did it "simply to keep the EPA off their backs.

"I don't think anybody planned to enforce it (the standards)," Reavis added. "The EPA came up with all these concentrations that you can't measure with any current techniques."

By mid-February 1992, when the new EPA rule takes effect, thousands of industrial facilities and sewage treatment plants nationwide could be forced into upgrading their equipment to meet the new standards as their water pollution permits expire. Although there have been no specific estimates made to date, Reavis believes Nevada plants are likely to spend millions of dollars trying to meet the requirements.

"Somebody is going to spend some awfully big bucks to remove all these things from their waste streams—without any idea whether or not it's even effective," Reavis said. "For example, in many places around here, naturally occurring arsenic exceeds the level the EPA wanted us to adopt."

This situation can be easily understood by the nation's minerals producers, who have already been severely impacted by unreasonable, unrealistic and often unattainable EPA standards. For example, the agency has classified naturally occurring substances (such as sand and dirt) as pollutants when they are present in wastewater discharges from mining operations—even when they are already present in the waters adjacent to the operation in vastly higher amounts.

If this overly stringent rule is actually implemented and enforced, most of the nation's industries (particularly mining and minerals processing) would find themselves in noncompliance on the effective date—and thereby automatically be subject to some horrendous fines. The EPA's current philosophy is "fine 'em and jail 'em" without extending any effort to actually help any particular business to achieve compliance. The agency's enforcement officials have stated that they want to "send a message to all potential polluters" by imposing the maximum fines and penalties available under administrative law, for the purported purpose of ensuring compliance.

In a classic example of a bureaucracy gone mad, the EPA actually sets certain chemical standards so low that industry can only comply by shutting down—and even then they're not off the hook. Businesses then find themselves perpetually liable for cleaning up the environment to meet the same impossible standards—even if they were not directly responsible for the so-called pollution involved.

If that doesn't stand your hair on end, consider this: If *all* of the current EPA standards and requirements applying to *all* business and industry at the present time were enforced strictly to the letter of the law, *all* business and industry in the United States would be shut down almost immediately. And the agency is still frantically trying to make things worse by formulating and implementing a massive number of *new* impossibly stringent rules.

We all want a clean environment, but is completely shutting down the entire country the proper way to achieve this objective?

Author's Note: Much of the quoted material about Nevada and the Nevada Division of Environmental Protection appeared in a front-page article in the November 7, 1991, Reno Gazette-Journal, in an item entitled "Nevada to Fight New Limit on Toxins," by Courtney Brenn.

41
Nevada's Mining Industry

Author's Note: Most of the minerals data and production statistics were obtained from the Nevada Bureau of Mines and Geology and the Nevada Department of Minerals.

DESPITE THE GENERAL DECLINE in most metal prices and production over the past few years and the nation's economic recession, Nevada's mining industry has continued to show remarkable strength and stability. This is primarily a result of the tremendous expansion of the state's gold mining sector during the 1980s, during which time the state attracted worldwide attention as a prime gold exploration target.

The mining industry is now second only to gaming and tourism in terms of size, gross revenues and importance to the economy. The total value of Nevada's mineral production continued to increase at a rapid rate until the onset of the national recession, and the minerals sector has fully emerged as the state's largest basic industry. In addition, mining is also contributing a significant amount of capital to the local economies throughout the state, because most of the major producing mines are located in rural county areas. Many of Nevada's smaller towns are almost entirely dependent upon mining as their primary source of employment, tax revenues and overall economic activity.

At present, Nevada leads the nation in the production of gold, silver, mercury and barite. The state's production of $2.54 billion in nonfuel minerals (hardrock and industrial minerals) last year was exceeded by only two states—California and Arizona. However, the national and international economic recession resulted in a 6.7% downturn in the overall value of nonfuel minerals production between 1990 and 1991, and oil production decreased about 20% in value over the same period. Geothermal production bucked the trend and increased slightly.

Gold production continues to dominate the state's mineral development activity and, since 1983, gold has consistently accounted for more than 65% of the total value of Nevada's mineral production. By 1990, gold represented approximately 85% of the total value of the state's nonfuel mineral production.

Nevada currently produces over 60% of all the gold produced in the United States and about 9% of all the gold produced in the world. The state's gold production has made the U.S. the second leading gold producer in the world and a net exporter of the metal, thereby helping to reduce the nation's international balance of payments. Last year, Nevada produced over 5.8 million troy ounces of gold valued at $2.13 billion. However, this represented a 1% decline in quantity and a 3% decline in value from the all-time highs reached during the previous year. When considering the economic recession, increasing mining costs, and much lower average gold prices during this period, this slight downward dip in gold output and value is not considered to be significant.

At the beginning of this year, Nevada's published precious metals resources (including minable reserves and possible subeconomic deposits) amounted to a total of about 135 million troy ounces of gold and nearly 497 million troy ounces of silver. Both of these figures reflect a modest increase in reserves last year, with ongoing exploration activity finding more new precious metal reserves than mining operations removed during the year. Industry officials estimate that the existing ore reserves are probably sufficient to sustain the gold mining industry at current levels in Nevada for at least another 20 years, providing that prices do not undergo a significant decline and government regulations do not force closures of existing mines or prevent the opening of new mines.

However, the trend towards continually discovering more gold reserves than are mined each year is not likely to continue, especially considering the overall decline in exploration activity during the recent years. For example, the number of new mining claims recorded with the Nevada Bureau of Land Management State Office has decreased from a high of 83,389 in 1988 to 21,624 in 1991, which indicates a very sharp decline in mineral exploration activity since 1988. Causes cited for this decline include: relatively low gold prices, uncertainties regarding continued access to public lands, and increasingly stringent federal regulations.

Although only one or two Nevada mines were being operated primarily for silver (these mines have since shut down because of low silver prices), most of the state's gold mining operations produce significant amounts of byproduct silver. As a result, Nevada's mines produced over 19 million troy ounces of silver valued at nearly $76 million last year. However, this represented a 12% decline in quantity and a 30% decline in value from the previous year. Similarly, no mines in the state are now being operated primarily for mercury, but many gold mining operations are producing mercury as a byproduct metal.

Other metals produced in Nevada during 1991 include: copper, 6.1 million pounds; lead, 1.3 million pounds; zinc, 21 million pounds, and an unreported amount of molybdenum.

The most important industrial minerals produced in Nevada last year (in order of value produced) were: aggregate (sand and gravel); diatomite; lithium carbonate; lime, cement, gypsum; barite; clay; silica, and magnesite. The total production of aggregate last year was 23 million short tons, which was a 12% decline from the all-time high of 26 million short tons in 1990. The state's gypsum production decreased from 1.6 million short tons in 1990 to 1.4 million tons last year. A general slowdown in California and Nevada construction industries contributed significantly to the decline in production of both aggregate and gypsum. However, the state continues to lead the nation in total barite production, shipping approximately 385,000 short tons in 1991.

Nevada also produced 3.4 million barrels of oil valued at $51 million last year, which was 15% less in quantity and 20% less in value than in 1990 but greater than production in all previous years. Although no new oil fields were discovered last year, four new wells became producers in the established fields in Railroad Valley. There are still many prospective areas in the state which remain untested for oil and gas.

Nevada's commercial geothermal power production reached an all-time high of 895,000 megawatt hours' worth $76.5 million last year, which was a 1% increase in amount and 5% increases in value over the previous year. Geothermal power-generating capacity in the state increased from 132 megawatts in 1990 to 147 megawatts in 1991, primarily due to completion of a second generating plant at Soda Lake. Eight additional geothermal power plants with a gross capacity of 183 megawatts are scheduled to be online by 1996, which would raise the state's total capacity to 330 megawatts.

Overall, the total value of Nevada's mineral, petroleum and geothermal production has increased from about $500 million in 1981 to approximately $2.68 billion by 1991.

Efficiency and technology are two of the most significant factors that have made mining in Nevada what it is today. The size of modern mining operations and equipment makes it possible to mine and process large volumes of ore with a high degree of efficiency. As a result, productivity has increased even as average ore grades have decreased, which is indicative of the rapid improvements taking place in the state's modern mining operations. This has made it possible, in many cases, to mine areas that were previously considered uneconomical.

Even with the increased efficiency and improved technology, many exciting challenges await Nevada's miners in the future.

Afterword

Dave W. Parkhurst's battles with his anti-mining adversaries ended with his sudden death from a massive heart attack September 16, 1993. Condolences and posthumous tributes from friends and colleagues soon poured in. Among those who paid tribute to Dave was Nevada Congresswoman Barbara Vucanovich, in her remarks on the floor of the U.S. House of Representatives three weeks after Dave's passing.

TRIBUTE TO DAVID W. PARKHURST

HON. BARBARA F. VUCANOVICH
OF NEVADA
IN THE HOUSE OF REPRESENTATIVES
Tuesday, October 5, 1993

Mr. Speaker, it is with much sadness I must report the recent passing of a constituent of mine, Mr. David W. Parkhurst, of Carson City. A prospector and mining consultant, Dave was the voice of the "small miner" in Nevada and the west. Dave was a long-time member and president of the Nevada Miners and Prospectors Association. He lobbied the state legislature on issues critical to the survival of individual miners and prospectors. Dave wrote monthly for the widely read *California Mining Journal* and also penned an informative column in the Nevada Mining Association's *Bulletin* styled "Parkhurst's Nuggets."

Through these articles Dave became a well-known correspondent and spokesman for small mining entities. He reported to his readers the workings of the Congress to reform the mining laws applicable to the public lands. Dave encouraged them to become engaged in the ongoing debate that threatens to force the individual prospector and "mom and pop" miners from our western public lands in their search for minerals that society wants. Few knew better than Dave Parkhurst the role the "small miner and prospector" have played in discovering and evaluating mineral deposits later worked by larger corporate miners.

It is with some irony, Mr. Speaker, that Dave's sudden passing occurred but a few weeks after mining claimants on the western public lands were called upon to pay "holding fees" for the first time, in lieu of performing labor on their claims to develop the deposits. Dave had been a vigorous proponent of "small miner" relief from this new tax burden, and indeed some exemption language was enacted by Congress, albeit not nearly as wide-reaching as he had sought. Well, Dave knew better than Congress that the federal budget deficit cannot be balanced on the backs of public lands users. He foresaw the abandonment of a tremendous number of mining claims because of the tax, and he has been proven correct. Only one-fourth of mining claims that were of record only a few years ago remain in good standing today.

Mr. Speaker, we will miss Dave in Nevada. **He fought the good fight** [emphasis added] against those who wish to see the public lands put off-limits to extractive industries. Dave Parkhurst's legacy will be the words he has left us to ponder as we debate the proper stewardship policies for the public lands. I, for one, have heard you, Dave. I will be steadfast in my efforts to ensure a place for miners and prospectors on our public lands.

About the Articles' Author

The biographical information about the author of the *CMJ* articles reprinted in the Dave W. Parkhurst Mining Writing Collection has been covered elsewhere in the collection, so in this space I'd like to simply share with the reader a couple anecdotes and personal observations about Dave in his role as lobbyist for small-scale miners and prospectors at the Nevada Legislature in the late 1980s and early '90s.

One evening in the late spring of 1989, Dave was watching the local evening news on television in our living room. He had come home weary and disgusted after another long day of listening to, and countering, anti-mining diatribes and misinformation coming from hostile legislators and environmentalists. Assembly Ways and Means Committee Chairman Marvin Sedway of Las Vegas had introduced an anti-mining bill every day for a week in his "Bomb the Miners Week" vendetta. Responding to an image that appeared on the TV screen, Dave raised his hand high in a mock Nazi salute and sarcastically barked at the image, "Sieg Heil!" This legislator, Dave explained, had introduced a particularly onerous proposal, A.B. 208, which miners had nicknamed the "S.S. Bill" (after the Nazi storm troopers) for its heavy-handed, punitive, and insulting approach to miners and mining in Nevada.

Growing up in rough-and-tumble, shades-of-the-Old-West northern Nevada communities in the 1940s and '50s, Dave learned quickly to stand up for himself and fight back when attacked; he was undaunted by the hostility and bullying tactics of anti-mining politicians during committee hearings or in other venues. Fortunately, he and the other mining representatives were able to defeat all of Assemblyman Sedway's "Bomb the Miners" measures.

Another success of Dave's and his fellow mining advocates' occurred in 1990, when Acting Governor Bob Miller declared the week of March 12, 1990, "Mining Industry Appreciation Week" and presented representatives of the state's mining community with a plaque recognizing the industry's contributions to the state of Nevada. This reflected what appeared to be a 180-degree turnaround for the governor, who had kicked off the start of his administration in January 1989 with a State of the State address that pitted miners against the children of Nevada in a bid to impose a huge tax increase on mining.

The Miller administration's subsequent change of attitude toward the state's mining industry was in no small measure a result of Dave Parkhurst's efforts representing the Nevada Miners and Prospectors Association, along with those of his fellow lobbyists.

Dave, a man of reason and "just plain common sense," remarked to me on more than one occasion, "You can't deal rationally with irrational people." But when he found rational voices among his adversaries and a willingness to listen to reason and consider the facts, he attempted to work with those adversaries to further the cause he was engaged in. That cause, in this case, was helping to protect the ability of miners and prospectors, particularly the small-scale operators, to function in an increasingly regulated society and in an industry under constant, often vicious attack by environmental extremists, anti-mining interests, and politicians currying favor with them.

Ironically, Dave's willingness to work with mining's opponents at all apparently did not sit well with some members of the NMPA (of which Dave was a former president). At one of the association's monthly meetings in late 1991, Dave was given a plaque expressing NMPA's gratitude for his representation at the 1991 Legislature; so far, so good. But then some of the NMPA members not-so-subtly chastised him for what they seemed to regard as working too closely with the opposition. I was there at the meeting when this occurred and was stung by what seemed to me a shabby, ungrateful way to treat someone who had worked as hard as my husband had in their behalf. He had *always* represented miners and prospectors honorably and wholeheartedly. Fortunately, the rift that had been created between Dave and the association healed over the next year or so, after conciliatory gestures were made by the group's membership. We both remained involved with the NMPA (I was editor and publisher of their newsletter) until Dave's death.

The reaction among Dave's friends in the NMPA upon news of his sudden passing was captured beautifully in a tribute by one of its members, Sue DeChambeau, which included these words:

We lost our champion when Dave died suddenly on September 16. His legacy to us is a package of courage, determination and the spirit to never give up in the struggle to protect our private property rights and the freedom guaranteed us by the U.S. Constitution.

APPENDIX A

GOLD PLACERS AND MINERAL DEPOSITS: *Their Formation, Deposition and Characteristics* is the first of the four volumes that make up the Dave W. Parkhurst Mining Writing Collection. It encompasses four main topics:
- How, where and why mineral deposits form
- The formation and characteristics of placer deposits
- The nature, characteristics and concentration of gold and where to look for it
- Prospecting for valuable metals and minerals other than gold

THE BASICS OF GOING FOR THE GOLD: *From Prospecting and Exploration to Small-scale Mining Project*, the second volume of the collection, consists of articles on the actual mechanics and processes of prospecting and mining, primarily for gold. It is divided into six main topics:
- Searching for gold
- Preparing to explore and prospect
- Prospecting, sampling, and evaluation
- Recovering the gold and other values.
- Prospecting and mining miscellany
- Case studies and small mining projects

A* CRITICAL INDUSTRY UNDER ATTACK: *The Struggle to Preserve Metals and Minerals Mining Viability in the U.S. is volume three in the set. The three main parts of the book concern these topics:
- The issues, challenges and threats affecting U.S. mining in the 1980s and early 1990s
- Miners and mining vs. anti-mining extremism
- Defense of the 1872 Mining Law as amended

FIGHTING THE GOOD FIGHT: *Mining's Battle for Survival in the American West* is the last of the four volumes in the Dave W. Parkhurst Mining Writing Collection. It covers five main topics:
- The U.S. government vs. mining
- Miners and the U.S. Forest Service

- Implementation of the mining claim holding fee
- Nevada miners vs. the politicians and anti-mining factions
- Metals and minerals mining in the Silver State

The four volumes in the Dave W. Parkhurst Mining Writing Collection are available at Amazon.com and elsewhere. For more information about the books or about Dave and his work, please visit **www.PineNutPress.com**.

APPENDIX B

YEAR OF PUBLICATION IN THE *CMJ*

Article No.	Article Title in Alphabetical Order	Pub. Year
18	$100 Claim "Holding Fee" Deleted from H.R. 2686	1991
19	$100 Mining Claim "Fee" in 1992 Federal Budget	1992
16	$100 Mining Claim Fee Proposal Defeated	1990
29	Anti-mining Forces Focus on the Comstock Lode	1988
13	Author's Response to USFS Letters on Claimant's Rights Issue	1988
10	Clinton Launches Attack on American West	1993
9	Congress Is Killing America's Economy	1992
1	Economic Policies and the Mining Industry	1985
5	EPA Gives Placer Miners the Shaft	1988
37	Governor Honors Nevada Mining Industry	1990
6	Grassroots U.S. Mineral Exploration Declines Sharply	1991
2	Great Economic Sacrifice, The	1985
23	Locating and Recording New Mining Claims	1993
7	Miners See Light at the End of the Tunnel	1991
3	Mining: Between a Rock and a Hard Place	1986
35	Nevada Enacts Mine Reclamation Law	1989
15	Nevada Governor Opposes $100 Mining Claim "Holding Fee"	1990
36	Nevada Legislative Actions on Mining Issues in 1989	1989
33	Nevada Legislature Seeks More Mine Tax Revenues	1989
31	Nevada Mine Tax Battle Grows Stronger	1989
32	Nevada Mine Tax Increase Approved by Voters	1989
34	Nevada Mine Tax Issue Finally "Clarified"	1989
39	Nevada Miners and NDEP Working on Regulatory Changes	1991
28	Nevada Mining Tax Bill Signed into Law	1987
30	Nevada Politicians Target Mining for Tax Increase	1989
40	Nevada Will Fight Unreasonable and Unenforceable EPA Standards	1991
25	Nevada's Gold Mining Industry	1987
27	Nevada's Mine Tax Issue Still Undecided	1987

Continued

Article No.	Article Title in Alphabetical Order	Pub. Year
41	Nevada's Mining Industry	1992
26	Nevada's Proposed Gold "Fee" Opposed	1987
24	New Mining Claim Fee Policy Implemented	1993
17	OMB Resurrects $100 Mining Claim "Holding Fee"	1991
22	Small-scale Miners Get Hit with $100 Fee and Bonding	1992
21	Small-scale Miners Get the Shaft	1992
38	Small-scale Mining in Nevada	1990
20	U.S. Government Planning to Sacrifice Small-scale Miners	1992
8	U.S. Mineral Exploration Is Being Decimated	1992
11	USFS Mineral Examinations Violate Claimant's Rights	1987
14	USFS Proposal Would Bypass Mining Laws	1989
12	USFS Responds to Author's Article on Claimant's Rights Issue	1988
4	What Are Our Mineral Priorities?	1988

References

Amendments to 1872 Mining Law and Federal Land Policy Management Act of 1976 (FLPMA). Department of the Interior and Related Agencies Appropriations Act, FY1991. H.R. 2686.

Brenn, C. (Nov. 7, 1991). Nevada to Fight New Limit on Toxins. Reno Gazette-Journal, p.1.

Browning, E.R. (January 1988). Letter to the editor. *California Mining Journal*.

Carlson, D.W. (January 1988). Letters to the editor. *California Mining Journal*.

Clinton, W.J. (1993). *A Vision of Change for America*, p. 78. Executive Office of the President.

Definitions of common and uncommon mineral varieties. (Act of July 23, 1955). Stat. 367, chapter 375, section 3. USFS regulations.

Definitions of common and uncommon mineral varieties. 30 U.S.C. 611 of the mining laws.

Department of the Interior Appropriations Act for FY1993. https://www.govtrack.us/congress/bills/102/hr5503.

Department of the Interior Appropriations Act for FY1994. https://www.govtrack.us/congress/bills/103/hr4602.

Effluent limitations, EPA regulations. (May 24, 1988). Final rule. *Federal Register*, Vol. 33, No. 100. Clean Water Act.

Effluent limitations, EPA regulations. (May 1988). Scope of this rulemaking. *Federal Register*. Section II. Clean Water Act.

Effluent limitations, EPA regulations. Section 509(b)(1) and section 509(b)(2) of the Clean Water Act.

Harn, Ken (January 1988). Letters to the editor. Ed. Note. *California Mining Journal*.

Mandatory Entitlements/User Fees/OMB and CBO Scoring. FY1991 OMB Appropriations Act negotiated budget compromise. Table II.B.

Memorandum of Agreement between BLM and USFS (1957).

Miller, B. (January 1989). State of the State address.

Miller, B. (May 1990). Letter to Nevada's congressional delegation opposing $100 mining claim fee proposal.

Omnibus Budget Reconciliation Act of 1993. (H.R.2264). http://www.congress.gov/bill/103rd-congress/house-bill/2264.

Ore mining and dressing; print source category; effluent limitations guidelines, pretreatment standards, and new source performance standards. 40 CFR Part 440 (frl-3361-7). Clean Water Act. EPA regulations.

Parkhurst, Dave W. (December 1987). USFS mineral examinations violate claimant's rights. *California Mining Journal*.

Parkhurst, Dave W. (March 1992). $100 mining claim 'fee' in 1992 federal budget. *California Mining Journal*. p. 21.

Robinson, Don (January 1988). Letters to the editor. *California Mining Journal*.

Scope of this rulemaking. (May 1988). *Federal Register*, Section II. EPA regulations.

Sedway, Marvin. (February 6, 1987). Assembly Bill 161 (AB161). 64th Session of the Nevada Legislature.

Senate Joint Resolution 22. 64th Session of the Nevada Legislature.

Vucanovich, Rep. Barbara. (June 25, 1991). Congressional Record — House. Remarks on $100 mining claim fee proposal, p. 16174. https://www.gpo.gov/fdsys/pkg/GPO-CRECB-1991-pt11/pdf/GPO-CRECB-1991-pt11-8-1.pdf

Index

$100 mining claim holding fee
    adverse consequences and impacts of, 51, 81-82, 89, 98-99, 102-105, 109-110, 113-115
    and federal deficit, 83, 105
    as political tradeoff, 101, 107
    as revenue source, 44-45, 84, 97-98, 103
    bonding and financial guarantee provisions of, 108, 109, 110
    cash-only requirement of, 99, 102, 103
    counterproductive nature of, 81-82
    descriptions, provisions and analyses of, 85-87, 88, 93-99, 102-105, 107-109, 111, 112, 113
    disproportionate impact of on small-scale mining and prospecting, 102
    final version of in FY 1993 budget, 111-115
    implementation of as Claim Maintenance Fee, 121
    initial defeat of, 41, 83-84
    locating and recording new claims under provisions of, 117-119
    major flaws of in 1992 proposal, 97-99
    Nevada governor's opposition to, 81-82
    OMB proposal for in 1992 federal budget, 81-82
    regulations and requirements of, 117-119
    removal of from H.R. 2686, 89-90
    resurrection of in 1991, 85
$20 million "loan" to State of Nevada by mining industry, 135, 136
$20.5 million tax prepayment compromise with Nevada mining industry, 139, 140-141
$30 million "gift"/"grant" from mining, Nevada Assembly's proposed, 135-136, 140
$42.5 annual Nevada mining tax plan, assembly Democrats' proposed, 136
10-claim small miners exemption (waiver), 121
12.5% gross production royalty on hardrock minerals, 52
150% mining tax increase proposal in Nevada Legislature, 148, 151. *See also* S.J.R. 22
1872 Mining Law
    and H.R. 2686, 89-90
    efforts to abolish or "reform," 114
    proposed changes to, 51, 52
1957 Memorandum of Agreement between BLM and USFS, 59, 61, 63, 71, 72
1965 Clean Water Act
    enforcement action on "toxic" chemical standards of, 189
    means of establishing new regulations, 29
    standards for effluent limitations, 29
1987 Nevada legislative session and mining tax issues, 135-138, 139-142
1987-1989 Nevada state budgets, projected impact on of mining tax revenues, 135-136
1989 Nevada Legislature and mining tax issues, 151-157
1989 small mines production, 182-183
1989 State of the State address by Nevada Gov. Miller vs. mining, 168
1992 budget's claim fee proposal, 95-97. *See also* $100 mining claim holding fee

1992 Interior appropriations bill H.R. 2686, 41
1993 Clinton budget plan, influence of environmental extremism on, 51, 54
43 CFR 3809 regulations, BLM revisions to, 115, 118
455% mining tax increase, Acting Nevada Gov. Miller's proposed, 148, 149, 151-156

A.B. 65, 175
A.B. 155, 176
A.B. 161 Jim Dandy gold fee bill in 1987, Assemblyman Sedway's
  arguments and testimony of opponents, 131, 132-133
  biased conduct of hearing, 131
  challenges to constitutionality of, 129-130
  defeat of, 140
  inequitable aspects of, 131, 132-133
  introduction of and justification for, 129-133
  negative impacts of on Nevada miners and mining, 132
  primary supporters of, 132
A.B. 172, 154, 176, 177
A.B. 178, 176
A.B. 190, 155, 176
A.B. 208, 155, 176
A.B. 315, 176
A.B. 770, 163, 164, 167, 176
A.B. 815, 176
A.B. 850, 176
A.B. 866, 164, 176
A.B. 872, 139, 140
A.B. 958
  enactment and provisions of, 171-173
  legislative action on, 177
  major provisions of, 172-173
  S.B. 549 as trailer bill for, 172, 175
A.B. 961, 177
abandoned mines hazard abatement, provisions for in A.B. 958, 172

Adler, Nevada State Sen. Ernie on A.B. 958, 171
AIP 1, 177
A.J.R. 13, 154, 177
A.J.R. 21, 163, 167, 177
American West, Clinton budget's negative consequences to, 51-53
annual claim rental fee. *See* $100 mining claim holding fee
anti-development activism in rural Nevada communities, 143-144, 145
anti-mining
  factions, influence of on $100 claim holding fee proposal, 101
  groups' tax-and-control efforts in Nevada, 161
  legislative proposals in 1989 Nevada Legislature, 163, 175-177
  publicity in 1989 Nevada mining tax battles, 151, 156
assembly bills in 1987 and 1989 Nevada legislative sessions, mining-related. *See* individual bills, abbreviated

Baldrige, Commerce Secretary Malcolm on value of U.S. dollar, 15
barite
  industry, effects on of USFS regulations proposal, 76
  production in Nevada, 182, 193, 195
basic industries in U.S., weakening of in early 1980s, 17
Bergevin, Assembly Minority Leader Lou
  on Sedway's A.B. 161 gold "fee" proposal, 130
  vs. Clark County anti-mining assemblymen, 154
BLM (Bureau of Land Management)
  43 CFR 3809 regulations, revisions to, 118
  Handbook for Mineral Examiners, use of by mineral examiners, 63, 66
  mining claim estimates and statistics for 1988-1992, 43, 44

regulation 3891 regarding validity
examinations, 70
regulations regarding $100 mining
claim holding fee, 117-119
rulemaking process, 112
Bomb the Miners Week in 1989
Nevada Legislature, 154-156, 165,
167, 176, 177
bonanzas, silver in Virginia city, 143
bonding and financial guarantee
requirements of Claim Maintenance Fee legislation, 112
Browning, E.R. of USFS
Letter to Editor from, 63
response to by author, 70, 71, 72
Bumpers' bill S. 433, 41
Bureau of Land Management. *See* BLM

Callister, Assemblyman Matt's mining
tax proposal, 153-154, 163, 167.
See also A.J.R. 21
Callister's A.J.R. 21, legislative action
on, 177
Carlson, Denton W. of USFS
Letter to Editor from, 64-66
response to by author, 71, 72
children-vs.-mining campaign in
Nevada, 1989, 151, 156
Claim Maintenance Fee
provisions of and state's compliance
with, 121-122
locating and recording new claims
under provisions of, 117-119
See also $100 mining claim holding
fee
claimant's due process rights in mineral
examinations and mining claim
contest actions, violations of, 57-
61, 63
clean environment, extremist vs.
realistic means of achieving, 189,
191
Clean Water Act, EPA's unrealistic and
unreasonable standards under, 29.
See also 1965 Clean Water Act
Clinton Administration's 1993 budget
plan's impact on West's natural
resource industries, 51-53
Clinton's *Vision of Change for America,*
excerpts of mining-related
proposals from, 51, 52
common and uncommon varieties of
minerals, USFS proposal to
redefine, 73-77
common varieties of minerals,
proposed revisions to categories
and uses of, 73
Comstock Historic District in Nevada,
143-145
Comstock Lode
and anti-mining forces in the modern
era, 143-145
new mineral discoveries and
prospective mining on, 144-145
Convington & Burling legal firm's
opinion on A.B. 161's constitutionality, 129
Craigie, Chief of Staff Scott's support
for Gov. miller's proposed 455%
Nevada mining tax increase, 153,
160
cyanide
heap-leaching technology in gold ore
recovery, 125
use in Nevada and proposed
legislation regulating, 151, 155,
176. *See also* A.B. 178

Daykin, Frank's legal opinion on
Sedway's A.B. 161 gold "fee"
proposal, 130
DeFazio's bill H.R. 2614, 41
Department of the Interior
appropriations act, 1992, claim
holding fee in, 93-99, 101-105
appropriations act, 1993, mining
provisions of, 107-110, 111-115
notification of EPA's new regulations,
32
procedures required for mineral
examinations, 57
disseminated gold ore mining, 125-126

Division of Water Resources Well Drilling Regulations, Nevada mining's issues with, 185-186

Dodgion, Lew NDEP meetings with mining representatives in 1991, 186

domestic minerals industry, actions detrimental to growth of and risks to, 25, 27

drill holes, changes to Nevada reclamation regulations pertaining to, 186

due process, violations of mining claimant's in USFS minerals examinations, 58

economic policies of U.S. in early 1980s, adverse effects of on mining and economy, 13-17

effluent limitations, EPA's new standards for, 29-33

eminent domain legislation proposed to exclude mining, 176

Endangered Species Act, radicals' use of against natural resource development, 40

environmental activists
  attacks on mining, 39, 156, 199, 200
  influence of in anti-mining legislation, 47, 60, 77, 101, 144

environmental degradation, safeguarding against, 108, 109, 127

environmental extremism and anti-mining efforts, countering, 39-42

environmental protection mania, 39

EPA wetlands definition and unrealistic standards, 40

EPA's "toxic" chemical standards enforcement, economic impacts of, 189-191
  for wastewater, Nevada's fight against, 189-191

EPA's punitive approach to regulatory enforcement, 190

EPA's standards for effluent limitations
  miners' concerns regarding, 30-32
  questionable means of establishing, 29-30, 32-33

exploration and mining technology, impacts of newer on gold mining, 125-126

extremist environmentalism vs. natural resource industries and development, 47

federal deficit, government policies contributing toward, 7, 48, 49, 52, 95, 125

federal EPA, stringent and excessive regulation by, 189-191

Federal Land Policy and Management Act, changes to, 85-86, 93, 95

federal land use policies, changes to in Clinton's 1993 budget plan, 53

Federal Reserve Board, monetary policies of in early 1980s, 7-10, 13-17

Fields, NDOM Director Russ, 81

Final Rule, EPA's 50 FR 47982 on effluent limitations, , 33

FLPMA. *See* Federal Land Policy and Management Act

foreign minerals industries, subsidization of by U.S. government, 53

foreign money, reliance on for economic survival, 49

foreign trade deficit
  consequences of for U.S. mining in early 1980s, 13
  effect of on monetary policies in early 1980s, 7, 8-9

Forest Service, U.S. *See* U.S. Forest Service

FY 1993 Interior Appropriations Act, $100 claim fee provisions in, 107-110, 111-115

Gaston, Assemblyman Bob in support of proposed $30 million "gift" from Nevada mining, 135

General Mining Law. *See* 1872 Mining Law

global recession in early 1980s, cause of, 13

gold and silver resources in Nevada in

1992, estimates of quantity of, 194
gold mining in Nevada
  1987, state of the industry in, 125-127
  from late 1980s to early 1990s, 193-195
  tax increase proposals for in 1987, 135-138
gold ore, disseminated, 125-126
gold production in Nevada
  in 1989, 179, 180
  in early 1990s, 193
gold rush of the 1980s in Nevada, 125
Gourdie, NMA President Jim
  on 1989 mining tax ballot initiative, 159
  on Callister's A.J.R. 21, 163
Gov. Bob Miller's 1989 State of the State address, Acting Nevada, 147
grassroots mineral exploration
  decline of and reasons for, 35-37, 43, 104
  in Nevada, threats to, 148-149
grassroots mining activism and $100 mining claim holding fee, 82
grazing fee in H.R. 2686 on ranchers using public lands, 90
gross income tax on mining vs. net proceeds of mines tax, 160
gross proceeds of mining operations, Sedway's push for, 140
gross production royalty on mining, provision for in Clinton's 1993 budget plan, 52-53

H.R. 2686, 1992 Interior appropriations bill, 41, 89
hardrock mining
  and minerals, proposed gross production royalties on, 52-53
  importance of to Nevada, 81-82
  reclamation bill for, Nevada's first, 171. See also A.B. 958
historic districts, legitimate purposes of vs. anti-mining tactic, 143, 144, 145

import reliance for mineral commodities, projected increase in, 37
industrial minerals production in Nevada, most important and volume produced, 195
inflationary pressures, effects of on U.S. monetary policies in early 1980s, 7, 8-9, 10
Interior department appropriations act, 1992, 41, 89. See also H.R. 2686

Jim Dandy gold fee bill, 129-133, 140. See also A.B. 208
Joerg, Nevada State Sen. Charlie
  defense of Nevada mining, 152
  remarks on A.B. 958, 171

Las Vegas politicians vs. Nevada mining, 153
legislation via administrative regulation, attempts at by governmental agencies, 30
litigation, rampant use of against natural resources development and mining, 40
locatable minerals. See uncommon varieties of minerals

marketability test for valid mineral discovery, 60
May, Assemblyman Paul, 137
Memorandum of Agreement between BLM and USFS, 1957, 59, 61
metals and minerals, demand for and supply sources of, 26-27
micron gold, mining of in Nevada, 125-126
Miller, Nevada Gov. Bob
  anti-mining campaign in 1989, 151-157
  efforts against mining in 1989 tax increase battles, 159, 160, 164
  honoring of Nevada mining industry, 179-180
  letter from to Congress opposing $100 claim fee, 81-82

211

Miller, Nevada Gov. Bob *(continued)*
  mining deductions, efforts to eliminate, 167
  targeting of Nevada mining for additional tax increases, 147-149
mined lands, reclamation of in Nevada, 127, 154, 156, 171-173, 175, 176, 177
mineral development, encouragement of in foreign countries, 36
mineral discoveries, small miners' and prospectors' role in new, 36
mineral examinations conducted by U.S. Forest Service
  and mining claim contests, 57, 59, 70
  authority for via 1957 Memorandum of Agreement with BLM, 59, 61
  binding of Forest Service to same regulations followed by BLM, 61
  claimant's rights, potential for abuse of, 57-61
  miners' concerns about, 69-72
  purpose and objectives of, 63
  violation of claimant's rights, 57-61
mineral exploration and development in Nevada, decline of since 1988, 193
mineral exploration and development in the U.S.
  $100 claim holding fee, adverse impacts of on, 83, 84, 85-88
  adverse impacts of FY 1993 budget provisions on, 99
  decline in and effect of decline on projected $100 claim fee revenues, 43
  destruction of by regulatory overkill and environmental extremism, 36
  downward trend of in years 1988-1992 and reasons for, 43-45
  risks to with Clinton 1993 budget plan, 52
  role of in a healthy mining industry, 126
  sacrifice of for short-term gain, 84
  small miners' and prospectors' role in and contributions to, 181, 183
mineral resources development in mid-1980s' North America, challenges to, 19, 22-23
mineral resources, relationship of to national security and societal well-being, 25
minerals
  common and uncommon varieties of and USFS proposal to redefine, 73-77
  produced in Nevada in 1989, 181
  saleable and leasable vs. locatable, 73-77
miners, Nevada. *See* mining and miners in Nevada
miners' issues with Nevada regulations and statutes, efforts to resolve, 185-186
miners' rights, protecting in interactions with U.S. Forest Service, 66
mining
  cyclical nature of, 19-20
  importance of to civilization and the economy, 179, 180
  misconceptions regarding, 127
  mining activism. *See* pro-mining activism *and see under* anti-mining
mining and manufacturing industries, foreign competition with U.S., 8, 15, 16-17
mining and manufacturing sectors, sacrifice and depression of by U.S. government, 13-17
mining and miners in U.S.
  adverse effects of $100 claim holding fee on, 85-88, 93-99
  attacks on and threats to, 17, 35, 39, 40, 44, 45, 130, 147, 151, 156, 157, 164, 168, 200
  countering of attacks on and threats to, 39-42
  government policies and regulations hampering, 40
  negative impacts of Sedway's gold fee proposals on in Nevada, 132

negative perceptions and stereo-
types of, 143
prospects for improvement of
conditions for, 39-42
selective taxation of, 103, 108
targeting by U.S. government of
small-scale, 101, 102, 105
mining boom in late 1980s to early
1990s, effects on of declining
mineral exploration, 43, 44
mining claims
active in U.S., BLM estimates of in
1991-1992, 43
contesting of, U.S. Forest Service
regulations regarding, 70
holding fee proposals. See $100
mining claim holding fee proposal
speculative purposes, holding of for,
86, 88, 95, 99, 103
validity examinations: BLM 3891
regulations governing, 70; USFS
notification of pending, 60, 66
mining claimants' due process rights,
violation of by U.S. Forest Service,
57-61, 71
mining exploration, shift of from U.S.
to foreign countries, 36, 37
Mining Industry Appreciation Week,
designation of by Nevada governor,
179
mining industry, U.S.
abnormal economic conditions
affecting in 1980s, 19-22
economic policies affecting in early
1980s, 7-11, 13-17
in Nevada, targeting of by politicians
for revenue, 147-149
negative impacts on of new EPA
water standards, 189-191
mining legislation in 1989, summary of
Nevada Legislature's actions on,
175-177. See also individual
measures
mining matters, legislating via administrative regulations, 73, 74, 75-77

mining opposition to $100 claim
holding fee proposal, 93
mining patenting
moratorium on, proposed, 89, 90, 91
Reid's full-market value amendment
regarding, 113-114
Rep. Vucanovich's opposition to
moratorium on, 89
See also H.R. 2686
mining severance tax, 129, 136, 148.
See also "Jim Dandy" gold fee bill
and A.B. 208
mining tax battles
in 1987 Nevada Legislature, 135-137,
139-141
in 1989 Nevada Legislature, 151-157,
159-161, 163-165, 167-168
mining tax increase, mining industry's
support for, 135
mining's war for survival, 40
mining-related legislation in 1989
Nevada Legislature, summaries of
and legislative actions on, 175-177.
See also individual measures
minting of precious medallions and
bars in Nevada, legislation
authorizing, 175
modern mining, environmental
protection and, 127
monetary policies of U.S. and political
expediency in early 1980s, 15
Mother Lode Miners, letter from
regarding USFS mineral examinations policy, 67
multiple use of public lands, threats
to, 54
Murray, NMA President Joe
on A.B. 872 advance tax payment
proposal, 139
on S.J.R. 22 net proceeds of mines tax
proposal, 135, 136-137

NAC 445 water pollution regulations,
redefinition of "small-scale facility"
in, 186

NAC 534 reclamation regulations, changes in to drill hole requirements, 186
National Defense Stockpile, important reliance for certain minerals in, 26
natural resource development, environmental extremists' opposition to, 39, 40
NDEP. *See* Nevada Division of Environmental Protection
Neal, State Senator Joe, 151
net proceeds of mines tax in Nevada
   amending of Nevada Constitution, 135, 141, 154, 159, 176, 177
   capping of tax rate at $4 million, 164, 167, 176
   eliminating deductions in computation of, 140, 148-149, 151-152, 160-161, 163, 164, 167, 175
   increase in, 135, 136, 137, 139, 148, 151-157
   legislation concerning, 129-133, 135-137, 139-142, 147-149, 151-154, 159-161, 175, 176, 177
   prepayment toward by miners, 135-137, 140, 141
   severance tax on, 129, 136, 148
   Rep. Vucanovich's opposition to moratorium on, 89
   See also A.B. 161 Jim Dandy gold fee bill *and* S.J.R. 22 mining tax proposal *and* S.B. 61 companion bill to S.J.R. 22
Nevada Constitution, amending of. *See* net proceeds of mines tax in Nevada, amending of Nevada Constitution
Nevada Department of Minerals
   minerals production data from, 182
   responsibility for abandoned mines program, 172
Nevada Division of Environmental Protection
   efforts to resolve miners' issues with, 185-186
   legislation concerning in 1989 Nevada Legislature, 156, 172
   mine reclamation enforcement responsibilities, 172
   new directive to regarding Nevada miners and mining, 186
   policy shifts on mining regulations, 185-186
   vs. Federal EPA on "toxic" chemicals standards enforcement, 189
Nevada gold and silver production, 1989-1991, 179, 180, 193-194
Nevada gold mining, boom in and targeting of for more revenue, 161
Nevada Governor Bob Miller. *See* Miller, Nevada Governor Bob
Nevada Legislature's mining subcommittee hearings on Nevada miners' issues, 185
Nevada mined lands reclamation, 171-173
Nevada miners
   efforts of to remove protections for in state constitution, 167. *See also* A.J.R. 21
   testimony of in opposition to A.J.R. 21, 163
   vs. anti-mining actions in 1989, 168
Nevada Miners and Prospectors Association at A.J.R. 21 hearing, 163
Nevada mining activism during 1988 legislative session, 156
Nevada Mining Association
   ad sponsored by NMA in support of S.J.R. 22, text of, 141
   at 1989 Nevada Legislature, 171
   at Gov. Bob Miller's ceremony honoring Nevada mining, 179
   report on disposition of mining-related measures in 1989 Nevada Legislature, 175-177
   testimony of on $20 million advance payment to state, 135
   testimony opposing A.J.R. 21, 163
Nevada mining claims, trend in decline of 1988-1991, 43

Index

Nevada mining industry
  attacks on in Nevada Legislature, 140
  contributions of and importance to the state of Nevada, 139, 141, 193
  damaging tax proposals potentially affecting, 139, 140, 141, 161-165
  efforts to resolve issues regarding regulation of, 185-187
  gold's role in success of from late 1980s to early 1990s, 195
  impact on of OMB's $100 claim holding fee proposal, 81-82
  impacts on of 1989 legislative proposals, 151-157
  legislative proposals concerning in Nevada Legislature, 129-133, 135-137, 139-141, 147-149, 151-157, 159-161, 163-165, 167-168, 171-173, 175-177
  proclamation by Acting Nevada Gov. Bob Miller recognizing, 179-180
  state of in 1992 and outlook for, 193-195
  targeting of with tax-and-control legislation, 163-165
  tax increases on as course of least resistance, 165
  taxation of as political football, 135, 137
  taxes paid by, 152, 153
  trashing of by politicians and news media over Question 1 support, 165

Nevada mining's
  $20.5 million advance tax payment to state, 139, 140-141
  fight for survival in 1989, 156-157
  support of mining tax increase, 152. *See also* S.J.R. 22 mining tax proposal

Nevada reclamation requirements for small operations, simplified, 186

Nevada small mines, production and employee numbers for in 1989, 182-183

Nevada State Education Association's support for A.B. 161, 132

Nevada State Legislature, 1987 session of and Nevada mining issues, 129-133, 135-138, 139-142

Nevada State Senate Committee on Natural Resources, special mining subcommittee of, 185

Nevada tax study in 1987 legislative session, 137

Nevada toxic chemicals list, comparison of to federal EPA's list of toxic chemicals, 189

Nevada water pollution control regulations' "small-scale facility" definition, 186

Nevada, minerals produced by small mine operations in, 182

Nevada, mining tax battles in, 135-138

Nevada's gold mining industry, state of and outlook for, 125-127

Nevada's ranking as gold producer in 1992, 193

new mining claims, requirements for locating and recording of under Claim Maintenance Fee provisions in 1993, 117-119

news media, bias of against natural resource industries and development, 47

NMA. *See* Nevada Mining Association

NMPA. *See* Nevada Miners and Prospectors Association

nonfuels mineral production in Nevada in 1991, 193

North America's mining industry, status of and outlook for in early- to mid-1980s, 19-22, 23

NRS 519A reclamation statute, amendment to small mining operations requirements in, 186

OMB $100 mining claim holding fee proposal
  in federal budget, 1992, 93-99
  in FY1993 budget, 52, 107-109, 111

OMB $100 mining claim holding fee proposal *(continued)*
   revenue projections from, unrealistic basis for, 43-44
Omnibus Budget Reconciliation Act of 1993, changes to claim fee provisions of, 119, 121
overregulation of mining in U.S.
   and harmful effects of, 23, 25, 40, 42, 44, 47, 48, 104
   and regulatory excess, 36, 37, 40, 41, 42, 44
   role of in decline of mineral exploration in Nevada, 194
   unrealistic and unenforceable EPA regulations on mining, 189-191

particulate matter, classification of by EPA as conventional pollutants, 29
political expediency, sacrifice of small-scale miners for, 107
politicians, anti-mining actions of, 157
precious metals mines, prospects for future of in 1980s, 22
private property rights and takings issues, 42
pro-mining activism
   in initial defeat of $100 mining claim holding fee, 83
   in Nevada, 156, 157, 161
   need for continued, 37, 101, 105, 161
   successes of, 39, 40, 41, 42
prudent man test for valid mineral discovery, 60

Question 1 mining tax ballot initiative, 159, 175, 177. *See also* S.J.R. 22
Question 1, sliding scale provision of, 165

Rahall, Rep. Nick's support for patenting moratorium, 90
Rahall's bill H.R. 918, 41
Reavis, Dick and NDEP stance on EPA's water standards imposition, 189
reclamation bonding, BLM's, 113
reclamation of mined lands in Nevada
   as part of modern mining's business operations, 127
   enactment of historic laws concerning, 171
   legislative proposals regarding, 175, 176, 177
Redelsperger, State Sen. Ken on A.B. 161 gold "fee" proposal, 130
Regional Ad Hoc Committee of Forest Service and Small-scale Miners, 65
regulatory enforcement in U.S. and trend toward overkill, 35-37
Reid amendment to mining patenting laws in H.R. 2686, elimination of, 114
royalty and lease payments to Nevada small miners, 181

S.B. 7, 175
S.B. 61 companion bill to S.J.R. 22, 148-149, 152-154, 160, 161, 167, 175
S.B. 212, 175
S.B. 287, 175
S.B. 469, 175
S.B. 476, 175
S.B. 549 trailer bill to A.B. 958, 172
S.J.R. 22 mining tax proposal, 135-137, 139, 141, 148-149, 152, 153, 154, 160, 167, 175. *See also* Question 1 mining tax ballot initiative
S.S. bill with surprise inspections of mines (A.B. 208), 155, 176
school class-size reduction, proposed funding of with mining tax increase, 164
Sedway, Assemblyman Marvin
   Bomb the Miners Week legislation, 165, 167, 176, 177, 199
   Jim Dandy gold fee bill, 129-133, 140, 147. *See also* A.B. 161
   vendetta against mining, 129, 130, 131, 147, 154-155, 156, 167
senate bills in 1987 and 1989 Nevada legislative sessions, mining-related. *See* individual bills, abbreviated

Senate Joint Resolution 22. *See* S.J.R. 22 mining tax proposal

Shaffer, State Senator Ray's chairing of mining subcommittee, 185

Sheerin, Assemblyman Gary's A.B. 770 anti-mining tax proposal, 160, 163, 164, 167, 176

Sheerin's A.B. 770, Nevada mining's opposition to, 164

silver mining and production in Nevada, state of in 1992, 194-195

sliding scale provision in Question 1 mining tax ballot initiative, 165

small miner waiver under new Claim Maintenance Fee, 121, 122

small-scale miners and prospectors
 and large companies in Nevada, mutually beneficial relationship of, 181
 contributions of to mining in Nevada, 181
 disproportionate impact of $100 claim fee on, 102
 exemption of from $100 mining claim holding fee, 111
 impact on of adverse factors affecting minerals exploration, 36
 sacrifice of for political expediency, 101-105, 107

small-scale mining operations in Nevada, 181-183

Spriggs, Assemblywoman Gaylyn on mining opponents' legislative "blackmail," 136

State Environmental Commission in Nevada
 actions by on mining regulations, 186
 responsibility of for mining reclamation regulation, 172. *See also* A.B. 958

State of Nevada Employees Association's support for Sedway's "Jim Dandy" gold mining "fee," 132

Stevens, U.S. Senator Ted's amendment to $100 mining claim holding fee proposal, 111

strategic and critical minerals, import reliance for, 25, 26

strong U.S. dollar
 and monetary policies in early 1980s, 7, 10
 as trade barrier for U.S. mining and manufacturing industries, 13-16
 negative effect of on U.S. mining in early 1980s, 20, 22

strong U.S. mining industry, importance of, 27

subsidization of U.S. minerals industry, fact vs. fiction, 54

Super Dollar as defense against inflation in early 1980s, 14

takings issue and private property rights, 42

tax issues, Nevada mining, 129-133, 135-138, 139-141, 147-149, 151-157, 159-161, 163-165, 167-168

Total Suspended Solids (TSS) in discharged waters, EPA standards for allowable, 29

toxic chemicals standards for wastewater discharges
 federal EPA's, 189-191
 Nevada's, 189

trade imbalance, counterproductive efforts to address, 17

U.S. Congress
 actions threatening U.S. economy, 48
 and legislative actions hostile to miners, 101-105
 anti-mining actions of, 107-110
 exemption of from laws and regulations imposed on population, 47
 hostility of toward natural resource industries and development, 47

U.S. economy
 reliance on foreign sources, 14
 threats to by Congress and government bureaucracy, 47-49

U.S. Forest Service
  and miners, efforts to improve relations between, 69
  and small-scale miners, working relationship between, 65- 67
  and Western Mining Council, 64
  claimed discretionary powers of in mineral examinations, 58-59, 60
  mineral examinations. *See* mineral examinations conducted by U.S. Forest Service
  Pacific Southwest Region's Minerals Area Management, 64-65, 66, 67
  policies and regulations, need for required vs. discretionary application of, 69-70, 71, 72
  proposal to redefine common and uncommon varieties of minerals, 76-77
U.S. industrial strength, weakening of in early 1980s, 7-10
U.S. industries, competitiveness of in world markets in 1980s, 7, 8, 9
U.S. mineral commodities and raw materials, reduced demand for in 1980s, 7, 9, 10
U.S. minerals
  decline in production of due to $100 claim fee imposition, 99
  producers, severe impacts on of unrealistic EPA standards, 190
  production capabilities, negative forces affecting, 20
U.S. mining laws, bypassing of with U.S. Forest Service rules-change proposal, 73-77
U.S. monetary policy in the 1980s, effects of on mining, 7-11, 13-17
U.S.-made products, reduced demand for in early 1980s, 13
unannounced inspections of Nevada mines, proposed, 155. *See also* S.S. bill *and* A.B. 208
uncommon varieties of minerals, proposed revisions to categories and uses of, 73
utopian idealism, sacrifice of multiple use of public lands to, 51
utopianism and environmental extremism, 105, 107

Vento's bill H.R. 1096, 41
Virginia City, Comstock Era vs. modern era in, 143-145
Vision of Change for America 1993 budget plan document, Clinton Administration's, 51-53
Volcker, Federal Reserve Board Chairman Paul and monetary policies in early 1980s, 15
Vucanovich, Rep. Barbara
  in support of S.J.R. 22, 152-153
  on grazing fee increase proposal in H.R. 2686, 90
  on patenting moratorium proposal in H.R. 2686, 89
  role of in removing mining claims fee from H.R. 2686, 89

Western Mining Council and U.S. Forest Service, working relationship between, 65
Williams, Assemblywoman Myrna in support of proposed $30 million "gift" from mining, 135

Editor Contact Information

Susan Lee (Sue) Parkhurst is the publisher/editor of Pine Nut Press in Minden, Nevada. To learn more about the publications and services offered by Pine Nut Press, or to see more of Dave's writing or his project photos, please visit **www.pinenutpress.com**.

The print version of the Dave W. Parkhurst Mining Writing Collection may be purchased from Amazon.com. E-book versions of the volumes are planned and will also be available at Amazon.com.

www.ingramcontent.com/pod-product-compliance
Lightning Source LLC
Chambersburg PA
CBHW020642220526
45464CB00001B/254